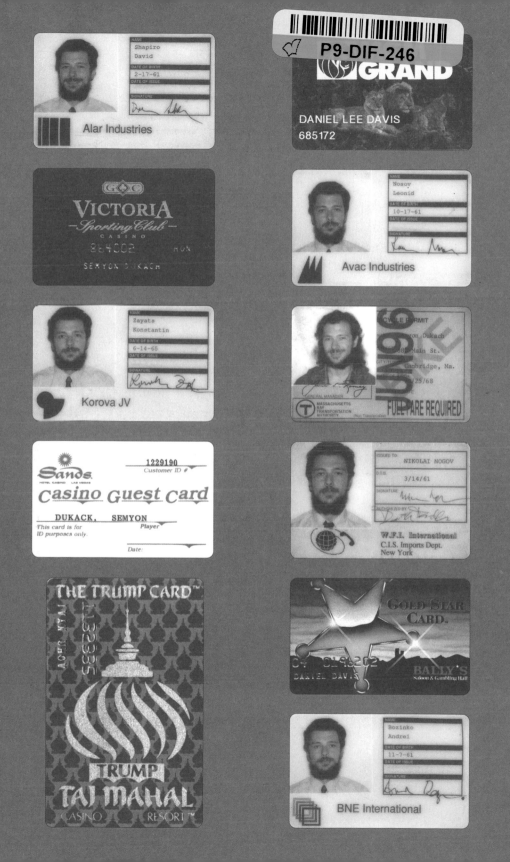

BUSTING VEGA$

Ben Mezrich

AFTERWORD BY
SEMYON DUKACH

BUSTING VEGA$

THE MIT WHIZ KID
WHO BROUGHT THE CASINOS
TO THEIR KNEES

WM

WILLIAM MORROW
An Imprint of HarperCollins*Publishers*

HarperCollins books may be purchased for educational, business, or sales promotional use. For information please write: Special Markets Department, HarperCollins Publishers, 10 East 53rd Street, New York, NY 10022.

FIRST EDITION

Designed by Jeffrey Pennington

Printed on acid-free paper

Library of Congress Cataloging-in-Publication Data

Mezrich, Ben, 1969–
    Busting Vega$: the MIT whiz kid who brought the casinos to their knees / by Ben Mezrich.—1st ed.
        p. cm.
     ISBN-13: 978-0-06-057511-3
     ISBN-10: 0-06-057511-5
     1. Cardsharping. 2. Gambling—Nevada—Las Vegas. 3. Gamblers—Massachusetts—Cambridge—Biography. 4. Massachusetts Institute of Technology—Students—Biography. I. Title.

GV1247.M495 2005
795'.092—dc22
  [B]   2005050616

05 06 07 08 09 WBC/RRD 10 9 8 7 6 5 4 3 2 1

AUTHOR'S NOTE

The events described here took place over the course of eighteen months in the 1990s. Some names and identities were changed to protect the innocent and not-so-innocent.

BUSTING VEGA$

PROLOGUE

*En Route from Harrah's, Atlantic City
Somewhere over New Jersey*

nder different circumstances, the moment might have seemed almost comic.

A forty-year-old Cessna four-seater airplane, lurching up and down in the turbulent dark. Two MIT kids dressed in velvet shirts and too-tight jeans, hanging on for dear life as they stared wide-eyed through the cockpit windshield. A plastic garbage bag full of hundred-dollar bills stuffed beneath their feet. And then the statement that hung in the rarified air between them:

"See, the thing is, I'm not really supposed to fly at night."

In the passenger seat, Semyon Dukach turned to stare at the young man sitting in the pilot's chair next to him. Victor Cassius was sweating. His mahogany skin was glistening and his thin black hair was matted against his forehead. His thin lips were pressed against his bright white teeth, and his eyes, usually narrowed into slits, were twin manhole covers. His collar was soaked through, and he was hunched forward over the steering yoke, his round shoulders bunched together beneath the electric purple shirt.

"What do you mean?" Semyon asked, his voice barely audible over the growl of the Cessna's twin engines.

"I shouldn't be flying after it gets dark. Instruments, and shit."

Semyon blinked. This was not good. Five minutes earlier, the sky outside the windshield had gone from a dull, gunmetal gray color to near blackness. Semyon could still barely make out the low cloud cover a few hundred feet below the small airplane, but beyond that, he couldn't see anything. No lights indicating towns, no geographic clues that could help them figure out their location. All he really knew was that they were somewhere over New Jersey. That twenty minutes ago, they had taken off from a small airfield outside of Atlantic City, heading due north.

"Okay, don't panic," Victor mumbled, reaching beneath the control panel. "Let me see. I know there's a switch down here somewhere . . ."

A moment later he found the trigger for the headlights. A tiny cone of yellow blinked against the sky. Great, Semyon thought. Like aiming a flashlight into a snowstorm. He tried to remain calm. He had been in dangerous situations before. He'd been beaten up. Held at gunpoint. Nearly thrown out of a hotel window. But somehow, this seemed worse. This was really his own fault.

Semyon couldn't believe he had let Victor talk him into climbing into the Cessna in the first place. The thing was ancient, and looked more like an old rusting VW Bug than an airplane. The bright yellow paint on the outside hull was scuffed and peeling, and there were visible cracks in the aluminum wings. The interior of the cockpit was cramped, all vinyl and plastic, and the poorly cushioned seats smelled like mildew. The control panel was like something out of World War II, bubbled glass gauges and black plastic switches.

But the derelict plane wasn't even the worst part of the equation. Right before takeoff, Victor had admitted that he'd spent the minimum required amount of hours with the flight instructor. Any more would have cost extra, and the whole point of the airplane was to save money. Victor had done the calculations; now that the team was hitting Atlantic City on a regular basis, it made more sense to buy a used plane to shuttle them back and forth from Boston than to fly commercial.

For Victor, it was always about the bottom line. He was always economizing, always optimizing. He had never paid full price for anything in his entire life. Two weeks into a relationship, he would move in with a girl to save money on rent. To describe him as cheap would be a disservice. He made being cheap into an art form. And when Semyon had pressed him on the subject of the expedited, money-saving flight lessons, Victor had replied in true MIT fashion. Though he had far less training than the average pilot, his IQ was far higher; Victor believed that IQ would more than make up the difference.

To Victor's credit, the flight down to Atlantic City had actually been quite pleasant. The weather had been perfect, and Victor had guided the Cessna down the Eastern Seaboard like a real pro. Right before the approach, he'd taken them low over the Hudson River, gliding around the Statue of Liberty so that the whole Manhattan skyline opened up in front of them, a gilded pincushion.

But tonight, it was a completely different story. Victor's last-minute decision to chart a course toward Princeton to visit an ex-girlfriend who was helping them recruit new team members had delayed their takeoff until near dusk. The wind had already been picking up by the time they'd reached the airfield outside of AC, and their takeoff had nearly given Semyon a heart attack. They'd almost been blown off the end of the runway, and the plane had climbed so slowly, Semyon had been certain they were going to crash into a row of town houses at the edge of the small airport. That they were still alive seemed like a miracle—and they were going to need a much bigger miracle to get back on the ground.

Victor tapped at one of the gauges with a stubby finger, then rubbed his eyes.

"We're okay. Nothing to this. But I'm going to need your help."

"I don't know anything about flying—" Semyon started. Victor cut him off.

"We need to triangulate our position. To find the airport in Princeton. You can use the radio frequency. There's a chart under your seat."

"You've got to be fucking kidding me," Semyon gasped. Triangulate their position? His PhD was in computer science, not engineering. He'd never triangulated anything in his life. He knew, in theory, how it worked. He was an MIT student, after all. But to find an airport in the middle of the night, while the plane was jerking around like a sick marionette—it seemed a terrifying task.

"It's either that, or we just set this sucker down in a field somewhere."

Semyon clenched his teeth, then bent forward to search beneath his seat. The plastic garbage bag took up most of the room between his heels and the ratty orange carpeting. He dug beneath the bag, feeling the thick bulges of banded hundred-dollar bills that pressed out against the cold plastic. He knew exactly how much was in the bag because he had counted it himself in the back of the rented car they had driven to the airfield. Three hundred and fifty thousand, thirty-five banded stacks of ten thousand. All of it taken from Harrah's in a little more than ten hours of play. Just the two of them, working in shifts, moving from table to table across the blackjack pit. The rest of their MIT team had watched in near awe from elsewhere in the casino as they had displayed their skills.

Semyon Dukach and Victor Cassius were the best in the world at what they did—although, to be fair, there were only about ten people in the world who knew their system.

Semyon and Victor were not card counters. In their minds, card counting was for jocks. Anyone could learn how to count cards by picking up a book or watching a special on television. You didn't have to be a genius to count cards.

What Semyon and Victor did was something quite different, something the casinos did not yet know about—something that nobody had ever put on paper. It was a system that had to be seen to be believed. It was a system that took true genius—and hubris—to pull off. The kind of hubris that made a twenty-four-year-old MIT grad student believe he could pilot a plane by means of his IQ. The kind of genius that could learn how to triangulate an airplane's position during severe turbulence.

His fingers trembling, Semyon finally found the chart beneath the plastic bag of money and yanked it onto his lap. Victor pointed him toward the radio receiver, and he went to work.

Angles and Dopplers and curvature patterns and radio waves.

Angstroms and meters and degrees. Semyon found himself entering that familiar zone of numbers and calculations. Although he'd never tried anything like this before, he'd spent most of his life studying one form of mathematics or another. The only thing that came easier to him than math was cards; and really, cards was just an extension of math. At its heart, this problem was all math as well. It was a matter of using the signals from two radio towers he could locate on the chart to calculate the position of the Cessna's own radio. An equation with calculable variables.

Five minutes later, he looked up from the chart and began directing Victor toward the coordinates he'd calculated. The Cessna shuddered as they navigated through the cloud cover, cutting hard into a tight descent. There was a loud whine as the tail rudder fought against the thick, humid air—and then, there it was. Flickering orange and red lights, a small, visible tower, and beyond that, a single paved runway. Victor let go of the controls long enough to clap his hands.

"Brilliant!" he shouted. "I always knew you were good for something."

"Just get us down," Semyon said, relief washing through him. His entire body was trembling, and he felt like he'd just run a marathon. He watched as Victor began checking the gauges. The lights of the small airfield became brighter as the Cessna's nose dipped to a forty-five-degree angle and they descended lower and lower. Semyon could make out the entire runway now, guessing it was at least fifty feet long. At the far end, he could see a parking lot of sorts, lined with dormant corporate jets and training airplanes, some feminine, curved and modern, some more masculine, angles and edges like the old Cessna itself.

They were about three hundred yards from the runway when Semyon noticed something that bothered him. Victor had gone still, his hands tight around the steering yoke.

"What's the problem?"

Victor cleared his throat.

"Nothing, really. Except, well, it looks like we've got a bit of a crosswind here."

Semyon peered out the windshield. He could feel the plane jerking with the wind, but he didn't see anything out of the ordinary.

"So?"

"Well"—Victor coughed—"landing a plane like this in a crosswind is kind of an advanced skill."

"What do you mean, advanced? You've practiced this, right? In flight school?"

Victor paused.

"Well, I did try it once. And I almost made it. The instructor didn't have to take over until the last few minutes. He said my approach was pretty good, though."

Semyon shut his eyes. His throat was constricting, and he could barely feel his fingers where he gripped the seat beneath him. They were going to die. He had trusted his life to a cheap maniac, and now he was going to die.

The Cessna dipped lower, curving deep toward the black strip of pavement. Muggy air rushed through the cockpit. Semyon could see that Victor was struggling against the wind, fighting to keep the nose of the plane aimed at the runway. Every few seconds the plane jerked to the right, then jerked back toward center. Closer, closer, closer. Semyon clenched his jaw, his eyes wide, his hands flat out against the console in front of him. And then there was a solid thud as the landing gear touched down.

"We're on the ground!" Victor shouted.

The entire plane shuddered, and Semyon felt as if someone had pushed him from behind. The tail was swinging back and forth with the wind. A second later, the Cessna coasted off the pavement and they were on grass, still rocketing forward.

"Get back on the fucking runway!" Semyon shouted.

But Victor was frozen. Staring straight ahead, his fingers white against the steering yoke. The plane continued down the grass, moving fast. Semyon looked at the speed gauge. Going down, but not quickly enough. Eighty mph, seventy, sixty. He turned back to the windshield, staring ahead with wild eyes. The runway lights were somewhere off to his left; here, on the grass, it was almost pitch-black. All he could see was the little cone of yellow from the headlight, bouncing along the grass, pitching up and down, illuminating shrubs and dirt . . . and planes.

Christ, the field of parked airplanes. Straight ahead, maybe another twenty yards.

"We're going to hit the planes—" Semyon started, and then he saw the brick wall. About three feet high, separating the field and runway from the parking lot. Semyon looked over, realized that Victor had seen it, too. There was nothing they could do.

"Hold on!" Victor yelled.

Suddenly the entire plane lurched up, and there was the sound of tearing metal. Everything seemed to crumple inward. Semyon slammed forward, his head slapping against the control panel. An instant, sharp pain tore through his right foot, and he screamed. Then he felt the heat. He opened his eyes and saw a huge ball of flame engulfing the cockpit.

Instinct took over. Semyon's fingers somehow found the searing metal of his seat-belt clasp, and a second later he was pulling himself across the passenger seat toward the door. His right foot was useless, shards of white pain ricocheting up from his toes to his ankle. Behind him, he could just barely see Victor kicking his way out of the wreckage on the other side.

His shoulder found the door, and he leaned into it with all his weight. There was a crack as the hinges snapped outward. His body tumbled to the grass. He rolled onto his knees, then started to crawl forward. His palms and cheeks stung, and he knew he was burned pretty badly. But the worst pain was still his foot.

He managed to get himself up on his good leg and hobbled as fast as he could, putting as much distance between himself and the burning Cessna as possible. Loud pops and bursts of heat erupted behind him. He could hear Victor shouting something from somewhere to his right, and he was glad his teammate had made it out as well. When he reached what he assumed to be a safe distance, he finally turned back toward the plane.

The fire was rapidly consuming the thing, orange flames licking up toward the black sky. The wings were curling inward like aluminum foil, and small explosions ripped up and down the tail. In a few seconds, the thing was going to be completely gone. Semyon could hardly believe they had both escaped the crash. There were sirens in the distance, and Semyon smiled, relief washing over him.

He saw Victor out of the corner of his eye. Running hard, wisps of smoke trailing behind him. No wonder he was running. It looked like his clothes were still on fire. Then Semyon came to a sudden realization. Victor wasn't running away from the wreckage. He was running toward it.

Semyon stared at his teammate in utter shock.

The money. Victor was going back for the money.

"That crazy mother—" Semyon whispered. He shook his head. Victor was insane. They had just narrowly escaped burning to death in the wreck. And Victor was going back for the cash. Three hundred and fifty thousand dollars. A good win, to be sure, but there was always more money to be won. There was always another casino to hit. It wasn't worth dying for, was it?

Semyon closed his eyes. The sirens were getting louder. In a moment, the ambulances and the fire trucks would arrive. By then, there'd be nothing left of the Cessna. Nothing left of the garbage bag tucked beneath the seats.

He took a deep breath, opened his eyes, and limped after Victor, back toward the burning plane.

CHAPTER 1

The Ranch
Carson City, Nevada

PRESENT DAY

ay too much velvet for three in the afternoon. Even for an oasis in the middle of the desert, a place that reeked of perfume and cigar smoke and Jack Daniel's, a place that I had been directed to by a cocktail napkin scarred by bright red lipstick. The velvet seemed to flow from everywhere at once; snaking down the wood-paneled walls, erupting from the low, tiled ceiling in voluptuous, pulsating waves, bursting from the shadowy corners, undulating beneath the plush daybeds and aging sofas that lined both sides of the ornate parlor. The stiff, blustery air-conditioning wasn't helping matters; the blasts of frigid air made the velvet dance and shimmer like living tissue. As a visual, it was more nauseating than enticing,

It had been a long taxi ride from the Strip, and I was dead tired from the heat outside. Arid, desert heat, not the kind that makes you sweat. Rather, the kind that cooks your brain. It was early September, and in this part of Nevada, that still counted as summer. You had to be crazy to come this far out into the desert in summer. Crazier still, to come to a place like this in the middle of a Friday afternoon.

I stepped deeper into the parlor, calming my nerves with deep breaths of frigid perfume, smoke, and whiskey. I wondered if the taxi was still waiting outside, as I had asked, or if the driver had simply pulled away as soon as I'd passed through the metal gate and made my way to the wire-screened front door. I certainly wouldn't have blamed him. Anyone deviant enough to pay three hundred bucks for the ride out to this ranch in the middle of nowhere deserved what was coming to him. I was no exception.

The truth was, this wasn't my first time in a place like this. For the past ten years, I'd traveled the world in search of stories, and sometimes those stories took me to places you really couldn't talk about at cocktail parties. Places like this ranch of paneled wood and velvet; a low, squat building that from the outside seemed to blend into the horizon—except for the neon sign on the roof and the decorative hitching posts in the driveway.

I took another step into the parlor, a circular space cluttered with anachronistic furniture, braced on one side by a long, mahogany bar. The bar stools were the same color as the velvet, a dusty crimson, and the sofas and daybeds had been upholstered to match. There were paintings on the walls, most of them of horses, a few of women and men in Wild West getups: hoop skirts, cowboy hats, boots with spurs. Kitschy, except I was pretty sure some of the paintings were authentic, since I knew that this place had existed, in some form or another, since the days of boots with spurs. An institution, of sorts, certainly more permanent than the neon behemoths of the Vegas Strip, built on something more primitive, seductive, and, indeed, human than the vice that had founded Sin City itself.

I'd almost made it to the middle of the room when I saw the woman sitting at the stool at the far end of the bar. Midfifties, short, squat, wearing a pink summer dress and white, high-heeled shoes. Her hair was a mop of curls, and her lipstick was an unnerving shade of orange. There was a glass of brown liquid on the bar in front of her. It could have been Coke, but just as easily whiskey. She heard my progress through the parlor and turned, but there wasn't any surprise in her gaze. I guessed that despite the heat and the time of day, this place still had its fair share of visitors.

She slid off her stool and turned toward me, smiling an orange smile.

"Welcome, stranger." She didn't seem to really look at me, instead focusing on a point just left of my ear. Seemed like force of habit;

maybe she didn't want to remember my face. "Have a seat on any of the couches, and I'll show you what we've got."

I lowered myself onto one of the daybeds, tucking my legs under the plush material. I was trembling beneath my white cotton button-down shirt, but the truth was, it was more anticipation than fear. Even though I was there for different reasons than the average client, the thrill was impossible to ignore.

The woman leaned back against the bar and clapped her hands. Then she cleared her throat.

"Ladies from the right!"

There was a shuffling sound, then a door opened along the right wall of the parlor. The first woman who came through the open doorway was ridiculously tall, maybe six feet, and her eight-inch stiletto heels made her seem almost gargantuan. She had flowing blond hair, glowing strands twisting down over her bare shoulders, gold rivulets dancing down the cavern of her surgically enhanced chest. Her bright red lingerie left little to the imagination. She was pretty, certainly, but more than a little terrifying as well. And she wasn't alone.

She was followed across the parlor by three more women, all in brightly colored lingerie and stripper heels. Two of them blond, one African-American. One of the blondes was short, a little more rounded, with a circular face and ovoid eyes. She could have been nineteen, if not for the spiderweb of lines at the edges of her overly pursed lips. The other blonde was much older, though she carried herself well. Surgery, again, and a lot of makeup, expertly applied in thick swatches across her cheeks, under her eyes, across her lips. The black girl was the only one of the four who was smiling, and it helped her stand out even more; she was by far the most beautiful of the girls, and she wouldn't have looked out of place on the pages of a magazine. Five nine, thin, smooth, brown legs, and a rounded, natural chest. Her outfit was lacy and white and fit perfectly over her curves.

It was a feat to pull my eyes back to the woman with the orange lips. She winked at me, then cleared her throat again.

"Ladies from the left!"

The door to my left had already swung inward before she'd finished her command. Four more girls entered the parlor. Two more
blondes, a brunette, and an Asian. Again, lingerie, again high heels
treading across the velvet sea. The blondes looked like sisters; matching green bras, panties, and garters, matching green eyes, matching,
egg-shaped boob jobs. The brunette seemed to be Eastern European,
with dark slashes for eyes, a sharp, upturned nose, and a jaw that
could cut glass. The Asian girl had her hair back in a severe bun, and
her outfit was all black leather and silver studs. She was playing the
part, her lips curled down at the corners, her charcoal eyes smolder-
ing with faux anger.

"Take your time," the lady of the house woodenly encouraged,
obviously reciting by rote. "And take a good look. The finest ladies in
the business. Money-back guarantee. Make your selection whenever
you're ready."

She glanced at her watch, elbow on the bar, chin resting in the
crook of her palm. I wondered how long the average customer sat in
the parlor, ogling the girls lined up in front of him. Did most of the
men who came here have something particular in mind, or was it a
point-of-purchase kind of business? In Amsterdam, the girls wore
gowns and stood on a raised stage behind the bar. In Tokyo, the
clientele sat in small booths, and the girls paraded through, one at a
time. In Vancouver, like Bangkok, you chose girls from behind one-
way glass. In New York, it was all passwords, descriptions, and of
course, the Internet.

Yes, I had been through this many times before in the course of
my research, but it was never the same—every place had its own
characteristics. Even a hundred miles out in the desert, this was still a
Vegas thing. The lineup was a show of sorts, and these were show-
girls. The understated cowboy theme, the routine of the "madam"
and her girls, all of it choreographed in the way an amusement park
choreographs its attractions. In effect, the Ranch was just another
ride in the neon, adult Disney that was Las Vegas. And I was simply
another paying customer.

I took a deep breath, thinking of the cocktail napkin in the
pocket of my jeans. The name scrawled beneath the directions to the

Ranch hadn't come with a description. I pretended to look over the girls, scanning the skin, silicone, and smiles.

"I'll take Gina," I finally said, praying that the crack in my voice was resounding only in my own ears.

The madam raised her eyebrows. I hadn't identified myself as a frequent customer, and she certainly didn't recognize me. But it didn't really matter; I was a man with a wallet and I'd made my choice. She shrugged, snapping her manicured fingers. For the first time, I noticed that the nails matched her lips.

"Thank you, ladies. The rest of you are excused. Gina, take this fine gentleman upstairs, and show him the ropes."

A few of the girls had disappointed looks in their eyes as they shuffled back through the opposite doors. I didn't pretend their disappointment had anything to do with my geek chic appearance; I knew they were thinking of a Friday-afternoon score, a good start to the weekend. As the doors closed simultaneously with a whiff of mingled perfumes, only one girl remained behind—the African-American woman in white lace with the pretty smile. She approached my daybed, holding out her hand, and I stumbled to my feet. Her fingers were warm against mine as she led me past the madam toward another door behind the bar.

"Glad that's over with," she whispered as the woman with orange lipstick went back to her drink. Gina opened the door with her free hand, revealing a short stairway ascending to the Ranch's second floor. "We all hate the lineup. Boobs out there getting compared like melons in a supermarket. That sort of thing is the reason most of us left stripping. But I guess nobody's buying melons without checking out the competition, right?"

I laughed, letting her lead me up the steps two at a time, trying not to notice how long her legs were, or how tight and sheer the white material was that hugged her rounded curves. She seemed much younger up close, maybe twenty, twenty-one. Her perfume was delicate and flowery, and her skin was pretty much flawless, a caramel brown. I wondered—why her? Was she part of this story, or was she just of the moment, locational, a prop of the scene?

"Right about now," she continued, taking me down a long hallway lined with nondescript doors, "I'd be giving you the menu. Then we'd be haggling about prices. By the time we made it to my room, we'd have everything locked down, and you'd be all ready for the inspection."

I raised my eyebrows. My heart was beating fast as I kept pace with her. This girl could move. "The inspection?"

She winked back at me. "That's where it gets fun. You take out your cock and I look it over. Then I rub it down with alcohol and Bactine. All free of charge, honey."

Her bluntness seemed incongruous with her youthful appearance. What she was describing was equal parts titillating and clinical. And it was also fairly distinct. Though there were establishments like this all over the world, in the United States, the Ranch represented something totally unique. A legal brothel, regulated by the Nevada Health Department, servicing one of the few prostitution-legal counties in the country. Just over the line from Clark County, where Las Vegas was located, which was supposedly prostitution-free, the Ranch was the closest place where a Vegas-based tourist could buy sex, or whatever else he desired.

"But the Bactine rubdown's about all that's free here, honey," Gina said as we neared the end of the long hallway. "Two fifty for oral. Five hundred for a half-and-half. Seven fifty for two cups. And a thousand if you want to go around the world. But everything is negotiable. I mean, even though I know you're not here for me, maybe I can interest you in some fun?"

She pulled my hand to her chest, running the back of my fingers against her bare flesh. I felt a tinge of heat in my stomach. I reminded myself that I was here for a story, nothing more. I wasn't sure what it meant to go around the world, but I was pretty sure it couldn't be considered a travel write-off.

"I don't think my publisher would consider it a necessary expense."

She laughed as we reached a door at the end of the hall. There was a gold number in the center of the wood: 232. From the outside it looked like a motel room, but I knew from my research it was much more than that. Gina was a private contractor, and this was her office. She lived and worked in 232, for a tour of duty lasting a few months, maybe as many as six. In that time, she could make a hundred, maybe two hundred thousand dollars. Some of the higher-profile girls made even more. It was all highly regulated work; weekly visits from a doctor to test for STDs, monthly HIV screenings, consultations with stylists, makeup artists, even visits from therapists and tax experts. Ethics aside, in terms of professionalism and health standards, the Nevada brothels were a paradigm of the form.

Gina pushed the door to her room open and stood to one side.

"Well, if you change your mind, I'll be back in ten minutes. Hell, for a few hundred extra, your friend can watch."

Before I could respond, she waved me inside, shutting the door behind me.

Her room was much more sparsely decorated than I would have guessed. A bank of wooden dressers along one wall, white shag carpeting, mirrors on the ceiling, and a single, king-size bed in the center of it all. No pictures anywhere, no windows, no knickknacks. No real sign of her personality in the room, which made sense when I thought about it. This was a place of business.

"Gotta love a room with a view."

I saw him on the ceiling first, because that's where my gaze had settled, and he was smiling down at me, framed by a cloud of off-white pillows. I shifted my view to the bed, where he was lying, spread-eagled against her king-size sheets, arms crossed behind his head. He looked relaxed, completely unfazed by the strangeness of the location for our first meeting in nearly three months. But that's how it always was with Semyon; as far as I could tell, he was comfortable in any setting, a true chameleon. More than a character trait, it was a calling, one of the keys to who he had become.

He rolled off the bed as I crossed the room, and we shook hands. He was a few inches taller than me, but I probably outweighed him by a good ten pounds. Everything about him was angular and narrow, from his build to the shape of his face. He had high, Slavic cheekbones, a thin jaw, a narrow forehead. His smile—and he was nearly always smiling—had more than a hint of wolf to it, stretching a little too far back, showing a few too many sharp teeth. He was good-looking, not matinee-idol handsome, but a character actor, an Ed Norton type. When he spoke, the words came out fast, tinged with enough hint of a Russian accent to force you to listen carefully, to catch every word.

On first impression, he was very amiable. Even after much time spent together, I liked Semyon, but I wasn't sure I trusted him. There was something dangerous about him, and it wasn't just that smile. I had spent many hours with him in Boston, and I knew his background.

"Now we're in deep, aren't we?" he asked, sitting back down on the bed. With a flourish, he pulled a wrapped deck of playing cards out of his pocket and tossed it between his hands. "A hooker's bedroom in the middle of the desert. I guess it's as good a place as any to start. This is as real as Vegas gets, isn't it?"

I knew what he meant. At least this place was honest, as honest as the mirror on the ceiling. Unlike the Strip hotels, with their neon and buzzers and bells, all of it a disguise to hide the gambling at their core. Semyon was right, this was a good place to start.

"What did you tell the girl?" I asked. "Gina?"

"I didn't tell her anything. I just informed the madam what room I wanted, and how much I was willing to pay for an hour. For all they know, we're just another kinky duo out here looking for kicks."

He rolled the deck between his fingers, looking at me. "I'm surprised you've never been here before. I know you've spent a lot of time in Vegas."

"Vegas, yes. But never out here."

He grinned. He liked the idea that he was showing me something new. When he'd first approached me, six months before, I'd basically shrugged him off. Another MIT kid with a story about beating Vegas, another twist on a tale I'd already taken as far as I thought it could go. I'd assumed his story was one I'd heard before. But then, over the next few weeks, I'd begun to discover things about the Russian whiz kid, things that made me think twice about completely blowing him off. When I finally sat down with him to hear his story, well, it made me want to dig deeper. The more I uncovered, the deeper I dug.

Semyon Dukach had indeed gone to MIT. He was a mathematical

genius. And it was true, he knew how to count cards. But Semyon was more than just another MIT cardplayer.

He tossed the deck of cards to me, laughing as I nearly fumbled them to the shag carpet.

"That's because you never came to Vegas with me before. You've barely scratched the fucking surface."

At one point in time, Semyon was the most notorious high roller in Sin City, perhaps even the world. He had been known by many names, but the one that had stuck was as flashy as many of the personas he had taken on: the Darling of Las Vegas. A living legend. He had beaten the game of blackjack more than anyone in history, in a way that had never been documented.

Although he knew how to count cards, his system was a different animal entirely. He and his team of MIT students had made millions, many millions, hitting the casinos harder than anyone else in the world—and yet, to this day, nobody knew how he had done it.

"Beneath the surface," he said, still smiling. "That's where it gets interesting. The gray areas. See, card counting is black-and-white. A fucking monkey can count cards. What we did, well, you need to be a little smarter to pull that off. And the casinos, they didn't like us very much. Because we were winning, we were a real threat. More of a threat than anyone else. We were hurting them, and they knew it. So things got . . . tricky."

His smile dimmed as he moved to the far edge of the bed. There was a hint of fear in his narrow eyes as he thought backward. It was unnerving to see him scared. He wasn't the type to scare easily. He had grown up dirt-poor in the slums of Newark and downtown Houston. He had clawed and kicked his way to a first-class education. He had been in more fights by the time he'd reached MIT than probably anyone in the school's history. He had built one of the most legendary teams to ever hit Vegas, kept it a secret for nearly fifteen years. Now, nearing thirty-three, he was most likely a millionaire many times over. And yet there it was in his eyes, fear.

He ran a hand through his hair, then suddenly pointed to a spot on the floor, a slash of shag carpeting beneath where the bed met the wall. His voice turned dead serious.

"This is where I found his body. Right here, halfway under the bed. His arms were twisted behind his back, and his head, well—"

He stopped, looked at me, then shrugged.

No, this wasn't a story anyone had ever heard before. And this place, this brothel in the middle of nowhere, it wasn't just a good place to start. For Semyon Dukach and his team of MIT geniuses, this was the place where it had all come crashing down.

CHAPTER 2

MIT
Cambridge, Massachusetts

A poster on a hallway bulletin board.

Twelve square inches of white paper marred by poorly stenciled black ink, daggered to aging cork by a half-dozen brightly colored pushpins. Crumpled at the edges, creased and yellowed by too many fingers, nearly lost and overwhelmed by a sea of other makeshift advertisements bleeding out across the pockmarked board: yellow construction-paper parchments offering cheap guitar lessons; pink and blue index cards heralding last-minute apartment rentals; still-warm photocopied flyers scalping concert tickets; dot-matrix cuneiform listings spiriting intramural sports teams that nobody wanted to join. And there, in the middle of it all:

> MAKE MONEY OVER THE SUMMER.
> PLAY WITH THE MIT BLACKJACK TEAM.
> SATURDAY MORNING, APRIL 12. ROOM 262.

Epiphanies were supposed to happen in places of great drama: the peak of an ice-blasted mountain, the middle of a storm-tossed ocean, a desolate beach on some deserted, primitive island. Epiphanies were supposed to strike like lightning, a bright, nearly spiritual blast of illumination, a fiery dagger cleaving through billowing black storm clouds of ignorance.

At MIT, epiphanies worked a little differently.

Semyon Dukach wasn't sure what it was about the strange little poster that made him stop by the overloaded bulletin board. Maybe it was the word *money* in big block letters, the lack of which had dominated his thoughts in recent days as his rent neared due. Maybe it was something more complex, the sense of ennui that had slowly been building since his return to school life after seven months of a sort of "extended vacation," spent traveling the world with a backpack slung over one arm and a now ex-girlfriend dangling from the other. Or maybe it was simply the monotony of the Infinite Corridor itself, the seemingly never-ending hallway that ran, like a spinal column, through the massive central conglomeration of buildings that made up most of the MIT campus. Whatever the reason, he paused, reading the words through a second time—four of them standing out, glaring like the fluorescent lights above.

Make Money. Play Blackjack.

Semyon smiled, rubbing at his triangular jaw. He glanced over his shoulder to see if anyone else was somehow sharing his moment; but all he saw was that interminably long hallway, snaking back behind him all the way to Mass Ave. Eight hundred and twenty-five feet of unremarkable corridor, lined on either side by identical classroom doors, architecturally wondrous in the sheer audacity of its banality. Every kid who'd ever passed through MIT knew the Infinite Corridor intimately. For most of four long years, it served as background to the MIT experience, a sort of architectural white noise. Then, as if by miracle, twice a year, the faithful gathered at one end to watch a remarkable phenomenon. On two astronomically predictable days in January and November, the sun, in its inexorable passage, exactly crossed the axis of the Infinite Corridor, and the setting orb passed directly by the Mass Ave. entrance, shooting bright light down the entire length of the hall. The phenomenon was so spectacular that those who had witnessed it had given it a name: *MIThenge*, an analogy to Stonehenge.

For some reason, Semyon was reminded of those two miraculous days as he stood in front of the bulletin board. The truth was, since he'd traded his backpack for a computer terminal in the school's comp lab, he'd felt himself descending right back toward the emotional funk that had inspired him to leave school seven months ago. Back then, it was Columbia, not MIT, but the mixed sensations of boredom, the lack of direction, the overwhelming blur of green computer screens and flickering fluorescent lights had been too much for him. He hadn't really expected a brief trip around the world and a transfer to MIT to cure his growing sense of dissatisfaction with the choices he had made, but he had hoped to come back fresh enough to finish school for good, get out into the real world. Fuck, at twenty-one he was way too young for a midlife crisis. Even if he had gone through more shit in twenty-one years than most people went through in forty.

Deep down, as much as he hated to admit it, he knew he belonged at a place like MIT. He belonged in a computer lab—because he was good at it, fuck, he was probably better in front of a blinking green screen than anyone his age. But he also knew, deep down, that he needed more. He *wanted* more. Maybe, after everything he had been through, he *deserved* more.

"Make money," he read again. "Play blackjack."

Then he reached forward and ripped the poster from the board.

The minute he stepped into the crowded classroom on the second floor of Building 16, Semyon knew he had made a mistake. He wasn't sure what he had been expecting, but this confederacy of MIT stereotypes seemed more a nightmare than an answer to his funk. All this place needed was a bank of computer screens, and he might as well have been spending a Saturday morning back in the computer lab.

There were at least forty people in the room, seated in chaotic rows of plastic, institutional-style chairs. A good thirty of them were male, and more than half of them Asian; not surprising, of course, given that this was MIT. The few girls in the room were of the local variety; thick glasses, hair in ponytails, baggy jackets and jeans. The men weren't dressed any better; thick sweaters, baseball hats, jeans with torn knees, Timberland boots. Semyon couldn't tell for sure, but most of the kids looked like seniors, though a few were obviously younger, maybe even freshman. Some of the seniors he recognized from the computer lab, but most of the faces were unfamiliar. He hadn't been at MIT long enough to make many friends—actually, he wasn't really the type to make friends easily, anyway. It took a lot to get past his walls, and he hadn't made many exceptions. A few girl-friends at Columbia, a couple of acquaintances from high school, that was really it. He talked to his father about once a week, his mother during the holidays. During his seven months of travel, he'd missed a total of twelve phone calls to his answering machine.

He scanned the room again, seriously considering heading back to his dorm. He wasn't looking for an extracurricular, and that's what this was looking like. A math club with blackjack as an excuse. Too many bright-eyed, eager kids in college sweatshirts and baseball hats for this to be for real. Sure, maybe you could make some money using math at a blackjack table, but from the looks of this room, it wasn't going to be anything more than a few slightly lucrative college field trips. This crowd was going to hit casinos and get away with it?

Semyon was halfway back toward the Infinite Corridor when he cast a final glance toward the front of the room—and something caught his eye. There was a large, steel desk at the head of the chaotic rows of plastic chairs, squatting beneath a wide, empty blackboard. A lone figure was leaning casually against the desk; a young man with dark brown skin and even darker, thinning hair, tapping an oversize eraser against his knee. The man looked vaguely South American, maybe from Colombia or Brazil. His eyes were a little too close to-gether, and his nose was sharp and upturned, set above a pair of ex-ceedingly thin lips. He looked to be average height, but thin, maybe even a little on the waifish side, and he was dressed oddly for the Saturday-morning meeting: a pinstriped gray suit, with a wide jet-black tie. He was surveying the room with quick flicks of his narrow eyes, continually tapping that eraser—oblivious to the fact that it was leaving a small cloud of chalk dust on the material of his recently pressed slacks.

Semyon raised his eyebrows, intrigued. The man at the front of the room looked to be twenty-four or twenty-five—too old to be an undergrad, but too young to be a professor. Maybe a TA, more likely a grad student. Was he the one who had put up the posters and called this meeting? Unlike the carbon-copy MIT kids, there was something innately interesting about the guy, the way he scanned the room, the way his eyes seemed to take in everything at once.

Watching the man watching the crowd, Semyon came to a decision. He'd give this thing a chance. Fuck, the computers weren't going anywhere. He turned and searched the room for an empty chair. There was nothing open near the front—another sure sign they were in an MIT classroom—but there were a few openings by the back wall. Semyon chose the one next to the least objectionable-looking student, a tall, lanky kid with stringy, dirty blond hair hanging down into his eyes, wearing a tie-dyed red T-shirt and black cargo pants.

As Semyon dropped into the empty seat, the kid glanced at him, then gave him a thumbs-up.

"Thought you were gonna jet. I almost did the same. Ain't this the finest bunch of geeks you've ever seen?"

Semyon smiled. The kid held out a hand.

"Owen Keller."

"Semyon Dukach," Semyon responded. Owen's handshake was limp, and he smelled of cigarettes. He also seemed a little drunk, which was odd, considering it was Saturday morning.

"Can't blame you for wanting to ditch," Owen continued, brushing the locks of hair out of his eyes. "You don't look like you belong with this crowd."

"What do you mean?"

"You look—well, hungry."

It was a strange assessment. Semyon assumed he looked like everyone else. Maybe he didn't wear glasses or a baseball hat, maybe he didn't own any MIT sweatshirts or Timberland boots—but he was a geek to the core.

"Fuck," Owen continued, not really pausing to explain himself, "we all aced the math portion of our SATs, right? We all memorized pi to the tenth digit, just for shits. So what makes these fuckers so fucking smug?"

Semyon looked around the room. He didn't think the other kids looked smug. They just looked like . . . MIT kids.

"What I mean is," Owen said, "you don't look like you grew up in Weston."

Semyon laughed, finally understanding. Weston was a wealthy suburb, fifteen minutes outside of Boston. Many of MIT's brightest came from similar pastures, country clubs posing as towns, pretty little places bristling with prep schools and golf courses.

"Not exactly," Semyon responded. "I was born in Moscow. When I was nine, my family emigrated to Newark, New Jersey. When I was fourteen we moved to Houston. I don't think my family could have afforded to park our car in Weston—that is, if we'd been able to afford a car."

Semyon wasn't complaining; his past was simply his past. He didn't have a chip on his shoulder. He knew his accent would always set him apart, but the rest seemed little more than lines in an imaginary biography.

"Sounds like a fun childhood," Owen commented, leaning back so his chair touched the wall.

Semyon shrugged.

"In Newark, I was a Russian kid in an all-black grade school. I got beat up a lot. In Houston, I was a Russian kid in an all-black high school. I got beat up a lot there, too."

Owen whistled. "Fuck, I thought I had it bad being the only kid who could do long division in dumb-fuck, Indiana. That sounds pretty grim."

Actually, it was much worse than it sounded. What Semyon had left out was that his family had escaped Russia by way of an Orthodox Jewish charity that had come to Moscow to help "refuseniks"—Jews who had been denied visas because of their religion. The thing was, Semyon's family wasn't really all that Jewish. Ethnically, sure, but they didn't know shit about the religion, had never been to temple, never even celebrated a holiday. They were just looking for a way out.

When Semyon and his parents had arrived in Newark, they'd found themselves swallowed up by a black-hatted cult. Not just Orthodox; actual zealots, strict fundamentalists who believed it was their religious mission to "educate" the poor, misguided Slavs who had lost their spiritual selves to the Communist state. The unpleasant experience had culminated for Semyon when, at the age of ten, he was dragged to the local hospital by two rabbis and forced to undergo an "elective" circumcision.

But Semyon wasn't going to go into all that with someone he'd just met—even though Owen seemed cool enough.

"I got over it," Semyon said, shrugging. But Owen wasn't listening anymore; his attention had shifted toward the doorway. A girl had just strolled through, and most of the eyes in the room were following her progress across the classroom. Tall, slender, with a shock of messy blond hair and ivory, almost transparent skin. A classic beauty, even in jeans and a puffy blue sweater, with barely any makeup or jewelry, just tiny dots of silver in the lobes of her ears and a whisper of red on her full, rounded lips.

"Who the hell is that?" Semyon asked. "I don't think I've ever seen her around the computer lab."

"Allie Simpson," Owen said. She had taken a chair two rows up, blithely unaware that even the air in the room had frozen in her wake. "She's premed, but lives off campus. Keeps to herself. Rumor has it she's got a boyfriend at Harvard."

Semyon sighed to himself. A boyfriend at Harvard—probably some rich kid with a Porsche and a single-syllable last name. Doubtfully a Russian immigrant who fixed people's hard drives to pay his rent. Okay, maybe a little chip on his shoulder; but still, he could admire her from afar. He had grown quite adept at that.

After the staggering blonde took her seat, the room went silent, and the well-dressed man at the front finally rose from the desk. He placed the eraser at the rack at the bottom of the blackboard, then turned to face the gathered crowd. Arms crossed against his chest, he smiled, his teeth too white, too straight, and small like pickets in a miniature fence, and introduced himself.

"Victor Cassius, former student like yourselves. Welcome, you made the right choice by coming here today because you're about to hear something that could potentially change your life. Three little words that have made many of us rich—me included—and could one day make most of you rich, too. You know what those words are?"

He paused for effect. Semyon watched him, the confident way he strolled back and forth in front of the blackboard, the way his suit hung efficiently from his wiry, catlike body. Charismatic, certainly, but more than that, he seemed worldly, cosmopolitan, exotic, so un-MIT.

"Blackjack. Is. Beatable . . ."

And with that, Victor spun off into a discussion of the game that had brought them all together. Almost immediately Semyon found himself tuning out a bit, satisfied to watch the catlike man with the volume in his head turned down. The rest of the room seemed captivated by the speech, but in truth, Semyon already knew the high points; he was not what one would call a novice to the game. He already knew that blackjack was uniquely beatable, that the stream of cards coming out of a dealer's shoe made the future predictable, and thus, inherently predictable. Years ago, as a child in Newark, he'd come to blackjack entirely by chance.

Apart from the daily beatings he'd received at the hands of the gangs that ruled the school yard, he had little to occupy his time when he wasn't studying math. So like other kids in similar predicaments, he had turned to the religion of the young and gifted: video games. Specifically, the daddy of them all, Pac-Man.

By the time he was twelve years old, Semyon had mastered the game of Pac-Man, reaching the high score on every machine he could find in northern New Jersey. But he hadn't been satisfied being the best Pac-Man player in Newark. He had wanted to be the best in the world.

So being who he was, he had gone to the local public library and, using his broken English, had convinced the librarian to do a search for books on the subject. To her surprise, she had found a single tome on Pac-Man strategy—written by a guy named Ken Uston. After reading the work, Semyon had become curious and had searched for other books by Mr. Uston. As it turned out, Uston had written numerous books, but most on another subject.

Blackjack. Ken Uston was a legend of the game and had actually invented the technique that the notorious MIT team had later mastered: team play. In a book called *The Big Player,* Uston had documented an amazing career as a professional blackjack player. He had made tens of millions of dollars, had traveled all over the world plying his trade—and had died under mysterious circumstances in a hotel in Paris.

Semyon had devoured Uston's story. He had even taught himself how to count cards. And then he had promptly moved on, letting his skills atrophy. The thing was, at twelve he had been too young to go to casinos, too young to gamble—too young to make any money off the skills he had learned. Maybe it was the efficient Russian in him, but he'd always kept his dreams accessible.

Nearly ten years later, he was hearing the Uston story once again, through the words of a strangely engaging young man in an MIT classroom. And this time around, Semyon wasn't too young for casinos—or the money locked away in the casinos' famous cages.

As Victor finished his speech, returning again to his perch on the edge of the desk, Semyon could feel the heat in the room. Victor had won them all over, turned every one of them on, and it was just a matter of showing them where to sign. But as for Semyon himself, something in him was still holding back. He knew the facts: Blackjack was beatable, card counters had an edge by keeping track of the high cards, raising one's bet when the cards were good—but even with a team, it seemed such a grind. A good card counter earned 2 percent per hand; a team, maybe a little more. Was that as good as it got? Was that enough to change one's life?

Semyon watched as the students filed back out of the room, chatting excitedly with one another about the MIT blackjack team, about what it would take to actually join, about the testing process and the "checkouts" that were necessary to earn a trip to Vegas. They all wanted to go to Vegas. They all wanted to give it a shot. But Semyon wondered—was there something more? Did Victor Cassius have secrets he wouldn't share with all of them, that he'd hold back for a select few?

Semyon found himself at the front of the room, a few steps behind Owen. Most of the other students had cleared out—but a few remained behind, milling about the desk, asking questions, trying to catch Victor's attention. Semyon noticed that Allie was there, hanging by the blackboard, a red flush to her cheeks. She had been sold as well, and for some reason, that thrilled Semyon more than anything else.

Without thinking it through, he suddenly found himself approaching the desk. Victor noticed him coming through the small crowd and smiled.

"So, what do you think? You want to be part of our little scheme?"

Semyon felt like those narrow dark eyes were digging right through him.

"I'm definitely interested. But is this straight card counting? Or do you do some of the more advanced techniques. Shuffle tracking, that sort of thing?"

Victor let out a little laugh, obviously impressed by Semyon's knowledge of the subject.

"We do everything you've heard of. And a few things you haven't."

Semyon found himself inadvertently glancing past the young man, at Allie, who was now leaning against the blackboard watching the exchange. Semyon felt himself suddenly emboldened by her gaze.

"And if I join your team, how much money can I make?"

Victor's eyes narrowed even more, and his voice dipped low.

"Well, that depends."

"On what?"

Victor cocked his head to one side.

"Exactly how far are you willing to go?"

CHAPTER 3

Boston, Massachusetts

Newbury Street, two blocks from the park.

A posh vein of high-end stores and Euro-styled cafés running through the direct center of Boston's ultrahip Back Bay. A place where property values seemed to double every few feet, a baby SoHo meets Madison Avenue, a meta–New York where women in furs and icy gray-haired men in suits had been replaced by young Persian boys in Prada and Gucci and frosted, giraffelike girls who looked like they'd just stepped off the runways of Milan. A young street in a young city, a designer enclave set in a college town that was metamorphosing every day into a facsimile of the metropolis three hundred miles down the coast.

Semyon strolled quickly past the brand-name stores, his hands jammed deeply into the pockets of his jean jacket. Hunched slightly forward, his body closed and protected—not simply against the unseasonably cold breeze that swept in from the Charles River four blocks to his left, but also against the opulent world of this section of the Back Bay. He had never felt comfortable in this playground of the nouveau riche. The ex-pat Iranian kids driving by in their Ferraris and Porsches, the Italians, French, and Armenian students sitting outside of Armani Café, sipping strange little cups of brown liquid as they compared hair gels and Rolex watches—Semyon was a different kind of immigrant. He had very little in common with the superwealthy Europeans whose parents rented them lavish penthouse apartments in this part of town. This was a place of money, for money; a place for shopping and being seen, and Semyon Dukach had never really been interested in either.

This particular Friday afternoon, he wasn't on Newbury Street to add to his minuscule wardrobe or model his circa-1984 jeans outfit. He was here because of a phone call he had received three hours ago. A phone call from Victor C.

It had been a long month since the meeting in the classroom at MIT. A month of training sessions, biweekly trips to Foxwoods, the Indian casino in Connecticut, and long hours of lonely practice in his dorm room. He had dealt deck after deck to himself until he'd re-learned the art of card counting, the Hi-Lo plus-minus system that the MIT team used to keep track of the high cards in a blackjack deck. The concept itself had been simple to remaster; the more high cards still in the deck, the better your odds, and the more you wanted to bet. It was just a matter of watching the shoe, keeping track of how many low cards had come out, how many high cards were still to be dealt. But in practice, it was much harder than it sounded.

Every ten days, Semyon and the rest of the team had been subjected to tests—called checkouts—in simulated casino settings. Always under Victor's watchful eye, they had gone through hundreds of decks, calculating the ratios of low to high cards, estimating the amount of cards still to be dealt, weighting the spotting numbers and calling in the Big Players, proving that they could survive in a casino setting as an integral part of this division of labor. Proving they had the skills to manage their money and manage the cards, to play as part of a team.

And not just any team—a huge team. As far as Semyon could tell, Victor was setting up the biggest team of card counters the world had ever seen. He had recruited more than forty kids, most with mechanical-engineering or mathematics backgrounds, and was training them all in shifts, teaching them the skills of card counting and testing them in an almost continuous process of checkout, failure, and recheckout. Many of the kids were sharp, some were even geniuses—but Semyon was pretty sure he was a standout star in terms of counting ability. He had a nearly photographic memory when it came to cards—an extension of his math and computer abilities, but more a product of his drive. He wasn't treating this like a hobby, as many of the other kids were. Semyon didn't have time for hobbies. He didn't have parents paying his way through MIT; in fact, he was the one sending checks home to his father, who still lived in Houston in the same shit-hole apartment where Semyon had spent his high-school days. When Semyon wasn't studying for his classes or practicing blackjack, he was working late nights in the computer lab, making the few extra dollars he needed to eat, fixing broken circuit boards and reprogramming bugged-out software. No, blackjack wasn't going to be a hobby; it was either going to be lucrative, or it was going to be a wasted month of his first MIT year.

Moving quickly down Newbury Street, Semyon prayed that would not be the case. The sheer enormity of Victor's team of card counters made him nervous; he didn't intend simply to be a cog in a blackjack machine that benefited a few powerhouse investors at the top—which is what he figured a large team would end up necessitating. Many times in the past month, he'd considered quitting after a particularly grueling checkout—not because of the stress of the cards, or the complexity of the math—but because of the meat-market setup of the sessions, the never-ending line of geeky young men eagerly waiting their turn at the makeshift blackjack table, the steady stream of Asian kids and vision-impaired girls in oversize sweatshirts learning spotting techniques from photocopied manuals that Victor had written and handed out.

For the most part, Semyon had kept to himself during the training process. Other than Owen, he'd hardly spoken two words to anyone else on the fledgling team. Owen had become his main practice partner, and they'd met twice a week at a local bar to talk cards over chicken wings and cheap beer. He liked Owen, though he barely knew any more about the scruffy kid than he learned from his first conversation in the MIT classroom. Owen never seemed to go to any classes, never talked about his past, and didn't seem to have any real friends or social life beyond their card-counting sessions. He was a moody kid, prone to long silences and sharp words, but he also had a sense of humor and seemed to enjoy Semyon's company. He lived in a dorm on East Campus but never once had invited Semyon over. He wasn't a friend, exactly, but the closest thing Semyon had found since entering MIT.

As for Allie, the hot blonde from the MIT classroom, he'd seen her a few times in passing. He'd even been present for her first check-out session, and he'd been impressed by her ability with the cards and her grasp of the math. She'd made only two counting mistakes in the six-deck shoe, too many to get her rated for a Vegas trip, but better than most. Everyone failed the checkout the first time through; everyone, that is, except for Semyon. He'd been the only one to pass on the first try, stunning the room—especially Victor—who had just gotten through explaining to the gathered crew that it usually took three or four checkouts before one was ready for Vegas.

And it was all about Vegas. Every kid who filed through Victor's training sessions was thinking about Vegas, about what it would take to get there, about how long one would have to train to be good enough for one of Victor's team excursions. And Semyon had to admit, even he had been caught up in the neon dream. He wanted to know what it was like to walk through a casino with cash in his pocket, to sit down at a real blackjack table and count the cards, call in the Big Player, pass the signals, and watch the money flow. It was a dream they all shared, a dream that crossed ethnicities and class boundaries, that linked the rich kids from Weston with a poor Russian immigrant from the mean streets of Newark and Houston.

It was the reason Semyon had perked up when he'd heard Victor's voice on the phone. It was the reason his heart had raced when Victor had invited him to his apartment in the Back Bay for a special meeting that Friday afternoon.

Semyon reached the last corner before the park and took a sharp left, heading toward the Charles. He'd had to check a map when Victor had told him the address over the phone; the only reason he even knew the names of the streets in this part of town was from a tour he'd taken during his initiation week at the start of his MIT year. Actually, he had been surprised when Victor had told him about the apartment on the first block of Commonwealth Avenue. He'd heard that Victor was notoriously tightfisted, and it seemed out of character that he would live somewhere like this. Maybe blackjack had already been quite good to him.

Semyon reached the pretty, tree-lined street and strolled the last few yards, coming to a stop in front of a four-story town house. Reddish brown exterior, with wide picture windows on every floor and elegant, nineteenth-century details, from the angled roof to the rounded redwood door at the front entrance. There was a cobbled walkway, separated from the sidewalk by a freshly painted, wrought-iron gate.

The gate was unlocked, and Semyon moved quickly up the three front steps to the wooden door. There was a brass knocker halfway up the wood. Cold to the touch and heavier than it looked. Semyon could feel each knock reverberate through his arm.

A few minutes later there was the metallic sound of latches being unlocked, then the door swung inward. To Semyon's surprise, Allie stood in the open doorway, hand on hip, white tank top rolled up to reveal a sliver of porcelain skin above her too-tight jeans. She was smiling at him as if she knew him, which was odd; though they had been in the same room a few times, Semyon hadn't really believed that the gorgeous senior would have been able to pick him out of a lineup.

"Hey, Semyon. We were beginning to worry that you had gotten lost. I know how treacherous the Back Bay can be."

Semyon felt his cheeks flush. He'd never been particularly shy around women, and had been through his fair share of college girl-friends. But something about Allie made him exceptionally nervous. There was a confidence to her, deep-seated and solid, emanating from somewhere far behind her pale blue eyes. The confidence flowed through every detail of her, from the playful way her short hair spiked up above her brow to the oversexed style of her clothes. Semyon felt himself shrinking in front of her, and that made him suddenly angry at himself. He wasn't going to let her secure nature push the trigger to his own rooted insecurities. Just because he fixed computers at three in the morning didn't mean he couldn't talk to a striking blonde.

"Sorry I'm late," he said, leaning against the doorframe. "I had a little trouble finding a parking spot for my Ferrari."

Allie smiled obligingly, ushering him inside. He found himself in an alcove with marble floors and high ceilings. A hallway led deeper into the town house, lined on either side by overflowing bookshelves. There was a crystal chandelier hanging from the ceiling and bronze sconces affixed above the shelves.

"Pretty nice digs, eh?" Allie said, running a hand through her short hair.

"I didn't realize Victor was doing so well. Who needs Vegas when you've got real estate like this?"

"It's his girlfriend's place," Allie explained as she started down the hall. "Victor's just squatting here. She's out of town for the week, so he's made it blackjack headquarters."

Semyon followed a step behind her. He could hear voices coming from the end of the hallway, and there was the thick scent of pizza in the air. Though he tried to keep his eyes straight ahead, Semyon found it difficult to keep his gaze from drifting toward Allie's long legs.

"I'm sure she'll love the idea of a few dozen college kids turning her house into a casino."

Allie laughed. "From what I've heard, Victor's girls last about as long as his suits. Have you ever seen him wear the same suit more than once?"

Semyon smiled. It was true: Victor's suits were a running joke among the blackjack trainees. He'd gone through every shade of gray by the time Semyon had remastered the Hi-Lo. As of a few days ago, he'd moved onto the blues. The suits seemed strange, coupled with the man's obvious mathematical skills and MIT persona; he talked like a grad student, but dressed like a high roller.

As they exited the hallway, Semyon caught sight of the high-rolling MIT guru at the far end of a wide, opulently appointed living room, bent over a strange metallic contraption that looked like a primitive photocopy machine. The device, ludicrously balanced on an antique mahogany desk, seemed as out of place in the vast room—replete with Victorian-era couches, Oriental rugs, and oil paintings hanging within gold-leaf frames—as Victor's eggshell blue suit. To add to the visual cacophony, a few feet in front of Victor a huge young man in what looked to be a velour sweatsuit was spread out across one of the period couches, a pizza box open on his lap. Semyon was sure he'd never seen the man before because he'd certainly have remembered what amounted to a humanoid eclipse. It didn't help the jarring panorama that the man's face was mostly hidden behind a bushy red beard or that his stringy hair sprouted out in crazed locks from beneath a tattered Boston Red Sox baseball cap.

Victor looked up from the machine long enough to take in Semyon and Allie as they entered the parlor, then continued whatever it was he was doing, his thin fingers flicking at levers and knobs.

"Welcome, Semyon. If you want pizza, you better move fast—and keep your hands and feet clear of the human trash compactor over there. His name's Jake, by the way, and he's a third year at Harvard Law. But we won't hold that against him."

Semyon shuffled onto one of the Oriental rugs and nodded at the behemoth, who smiled at him. There was pepperoni in his teeth. Though it was hard to tell beneath all that fur and fat, his eyes looked young; he was maybe twenty-two, twenty-three years old. Probably extremely sharp, considering his pedigree, though Semyon wondered how good his math skills could be. Harvard Law wasn't exactly MIT.

"Okay," Victor interrupted his thoughts as a whirring sound was emitted from the machine. "And here we go."

Victor leaped up suddenly, snatching a small plastic object from within the metal. Then he crossed the room in quick, short strides. Semyon was a head taller than Victor, but the swishing material of Victor's dapper suit made up in style what he lacked in height. He stopped in front of Semyon and handed him the small slab of plastic. It was warm from the machine, as thin as a credit card, and clear at the edges.

Semyon turned it over in his hand and saw his own picture staring up at him, in the center of a craftily designed ID card. He recognized the photo from his MIT face book. Smiling, scruffy, his hair a bit wild, a few days of growth on his chin and around his cheeks. He had just returned from his seven-month world trip when they had stood him up against a bookshelf and snapped the shot for the welcome book.

Beneath the picture was a name, but unlike the photo, the name was not one that Semyon recognized.

"Nikolai Nogov?"

Victor smiled. "Gotta admit, it has a nice ring to it. If you don't like it, we've made about twenty more with different names; you can choose as many as you want. Consider this your swearing in. Congratulations, you are now officially a member of the MIT team."

Semyon's eyes widened. Just like that? He glanced over Victor's shoulder, at the machine on the far side of the room. Fake IDs. Twenty of them, all with his picture. He wasn't sure why, but the thought made his stomach tighten up. This was a first step, toward what he didn't know, but certainly a step. He was now a member of Victor's team. What exactly did that mean? And was it illegal to make fake IDs? He looked at the card again, but he didn't see anything that seemed to imply that it was federal- or state-sanctioned. At first glance, it appeared to be pretty innocuous, something a kid would make to try to buy cigarettes or beer. However, the idea of a machine that could churn out dozens of IDs was both exciting and disturbing.

"And what about the others?" Semyon asked, looking back at Allie, then at the large man on the couch. "Where's the rest of the team?"

Victor moved toward the pizza box and liberated a slice from the Harvard law student. "There aren't any others."

Semyon raised an eyebrow, confused. There were thirty other kids training every day, learning how to count cards. Surely Victor hadn't sent them all packing. Many of them had math skills that nearly rivaled Semyon's. Most of them were on their way to becoming experts at Hi-Lo.

"Well, that's not exactly true," Jake chimed in from between bites of pizza. "Victor likes to be dramatic. The others are part of a much larger scheme, which at the moment we're calling Strategic Investments. A card-counting team, pure and simple, with big players and spotters and gorillas and so on. But all of that is bullshit compared to what we're going to do."

Semyon felt Allie standing next to him on the Oriental rug, and he wondered how much of this she already knew.

"We're not part of the card-counting team?" he asked.

Victor crossed his legs, straightening the pleat of his pants. "We're something else entirely. Me, you, Jake, and Allie. That's the core of our merry band. When we hit Vegas, we'll bring in a few more support troops, carefully chosen from the peons crunching numbers."

Semyon thought of Owen and decided that when the time came, he'd push for his friend to join their core group. Although he barely knew the kid, at least he felt comfortable with him, which was more than he could say for anyone in this lavish parlor. Allie made him nervous because she was beautiful and way too confident; Jake, the bloated law student, was a complete unknown; and Victor, in his ever-changing suits, a sprightly, charismatic genius who had set up a fake-ID machine in some exchangeable girlfriend's Back Bay mansion, was as wild a card as an extra ace in a blackjack shoe.

Owen aside, Semyon's thoughts turned back to Victor's words, and one of them resonated, a flash of auditory neon.

Vegas. *When we hit Vegas.* So Semyon was really on his way. But now he wasn't even sure what he was going to be doing when he got there. Blackjack, certainly, but not card counting? What kind of core team was Victor attempting to create?

Victor seemed to read his train of thought and rose from the couch. He took a napkin out of his suit pocket and gingerly wiped the pizza oil from his fingers.

"Over the next two days," he said quietly. "Jake and I are going to teach you three techniques that we've mastered, improved, and made our own. These three techniques are the most powerful weapons to ever be unleashed in a casino—and once all four of us master them, we will literally change blackjack history."

Semyon felt his throat go dry. He didn't really know what Victor was talking about, but he could feel the tension in the air, the way the room seemed to tighten in around his words. In the classroom at MIT, Victor was a ringmaster, guiding all the attention toward himself with verbal ease. But here, in these small quarters, the import of his words seemed almost religious in nature. He really believed what he was saying. When he looked at Semyon, straight in the eyes, it was like propane going off, a heat strong enough to ignite a fiery crusade.

"With these three techniques, we're going to make millions. We're going to bust Vegas wide open. And the casinos will never know what hit them."

CHAPTER 4

MIT
Cambridge, Massachusetts

Please excuse the beige. Somehow, the future of the world always seems to start with the color beige."

I laughed, not simply because the frumpy kid's quirky sense of humor was finally breaking me down, but because he had a point. I'd found that often the most momentous situations arose out of the most banal of settings. This third-floor hallway in a rectangular office building on the MIT campus, with its beige walls, beige ceiling, and beige carpeting, was simply another case in point. When one of the walls changed to glass, revealing a wide area of open desks, computer terminals, printers, fax machines, and electronic workstations, the switch in architecture barely registered. This was a nondescript place in a nondescript building on a campus that often lacked description. At the same time, this was also one of the most important places in the world. Inarguably, this was where critical advances in science, technology, and culture were finding root, a place that would literally change life on planet Earth in ways most of us could not yet comprehend.

"Welcome to the MIT Media Lab," my guide continued, with an awkward flourish that took both hands and most of his narrow shoulders to pull off. "It's not much to look at, but hey, here the fun never stops. And the company, well, it's pretty much enough to make you want to kill yourself."

Evan Stein wore a fairly crazed grin as he held open a door set into the glass wall. He was a strange-looking kid, from his wild mop of frizzy brown hair to his deep-set, hangdog eyes. His jeans were way too baggy, hanging down over a pair of scuffed yellow high-tops. He was wearing a backpack on both shoulders, and I could see the tip of a skateboard peeking out through the green vinyl.

Evan was in the second year of his PhD program at the Media Lab, though if I had to guess based only on his appearance, I would have pegged him to be still in his late teens. He had been referred to me by Semyon, and at some point in the past they had worked together on a computer enterprise, something that had initially been funded by Semyon's blackjack money, but had become something much more, an Internet company that only a few years ago you could read about in *The Wall Street Journal*. Over the past six months, the Internet company had disappeared, but Evan Stein was still kicking around MIT, earning another degree while awaiting the next money-making opportunity that would free him from the books and the beige. In Semyon's opinion, with Evan's intense mind and his even more intense lust for commercial success, he was a perfect fit for the infamous Media Lab. A genius, but beyond that, an independent mind, a wild thinker—and a hungry soul.

"You've been here before, right? So you know the basics."

He led me past a pair of Mac computers to a steel bench over-loaded with wiring and circuit boards. Above the bench, affixed to the wall by two plastic hooks, was what looked to be a down jacket—except this jacket was right out of a sci-fi movie. Day-Glo wires ran up and down the sleeves, leading to small diodes and blinking switches at the cuffs, collar, and central zipper. A plasma readout was attached below one lapel, brilliant green numbers indicating temperature, pressure, humidity, wind speed, and a dozen other variables that made little sense to anyone uninitiated to the world of cyborg clothing.

Evan was right: I had been to the Media Lab before, researching an article for some glossy magazine; things like the wearable computer on the wall in front of me did not surprise me, because I had learned that the Media Lab was all about surprises. Around every corner lurked technological delights that ranged from the mildly ab-surd, such as a jacket that interacted with its wearer, to the com-pletely bizarre. On the floor below us, chemistry grad students and mechanical-engineering professors were busily working together on a purely chemical computer, one that ran on the reactions of minute liquids, built with zero solid components. On the floor below that, engineering students and ceramic experts were working on the kitchen of the future, complete with pots and pans that communi-cated with one another, a stove that knew what it was cooking, even a refrigerator that kept tabs on what it was keeping cold, predicting what its user would want to eat next.

Machine shops full of robots, enormous lasers, printers that printed circuits onto circuit boards instead of ink onto paper, the Media Lab was a wonderland of high-tech thought, a cross-disciplinary marriage of art and science. Founded in 1982 by one-time architect Nicholas Negroponte and a team of interdisciplinary scientists, the Lab had grown to include forty full-time staff, eighty visiting professors, and nearly a hundred and fifty postgraduate stu-dents. It was a huge endeavor that was also a collision of industry and study funded almost entirely by corporate sponsors—who got first dibs on the tech that was developed in return for their multimillion-dollar investments—a large part of its purpose was to turn creative, wild thought into commercial endeavors. A perfect place for a kid like Evan Stein, and an even better fit for Semyon Dukach.

"So you met Semyon here," I asked as we wound our way through a group of students hovering over a batch of freshly minted "smart" LEGOS—another Media Lab invention. "You were both students of the Media Lab?"

Evan shrugged. "Semyon was in and out of here all the time. He was sharp as hell at computers, but he wasn't satisfied poking around circuit boards or designing supersmart refrigerators. Still, we got along pretty well. He understood what was going on here, but de-cided to go in a different direction."

I watched as a pair of Indian kids played catch with some sort of robotic Slinky.

"What did he understand?"

"It isn't enough to be smart," Evan responded. "You've got to be street-smart. There's got to be a bottom line in everything you do. There's no point to math and science if it doesn't have a real-world application. And there's no application more real than the art of making money."

He stopped in front of a table cluttered with what looked to be stuffed animals. Bears, badgers, monkeys, dogs. When he waved his hand over them, they all seemed to come alive, barking and yapping and whirling around. Then he snapped his fingers, and they all went dead.

"Nerds like us, as a species, are anarchists at heart. We're all for subverting authority and dodging the rules. But we're also about working the system to our advantage. Finding ways to use our brains to turn things our way."

He turned to face me, wearing that grin again.

"Did you know that hacking was invented here, at MIT? The whole hacker culture comes from here. Originally, hacks were just elaborate practical jokes. Pranks, the more technically difficult, the better. The most famous one involved a campus police car. A group of seniors managed to take one apart and reassemble it on top of the dome above the student union. Another hack involved an inflatable balloon that launched itself at the fifty-yard line in the middle of a Harvard–Yale game. Hacks, you see, are a way of subverting authority, but doing it in a way that affirms your own value, your own sense of superiority."

He was summing up the MIT character in succinct terms, and my thoughts turned back to Semyon. The concept of hacker culture came to mind—anarchistic, rebellious, more than a little dangerous. The team that Victor had put together, the three techniques he had taught Semyon and the others—they were akin to a great, incredibly profitable hack aimed at the casinos in Vegas and around the world.

Card counting; that was a very different thing. Card counting was legal. It was simply using one's brain to beat the house. Black-and-white. But Victor's three techniques, well, that was more of a gray area. Because the three techniques weren't simply playing within the rules of the game, they were a manipulation of the elements that made the game work. Was it really legal? Was it cheating? Or was it something in between?

In my mind, what Victor taught Semyon was really a form of mathematical sorcery. A marriage of art and science, like what went on here, at the Media Lab, where Semyon had spent his time when he was not engaged in Victor's scheme.

"See, the thing is," Evan said, turning one of the stuffed monkeys over in his hands to show me the circuit board that made it tick. "We have reason to feel superior because we understand something most people don't. Deep down, it's all about math. Math is the key to everything. The most powerful weapon anyone could ever have. Math makes bombs. Math makes guns. Math makes monkeys that leap when you walk by. Math has real-world applications, and it's guys like Semyon and me who have the knowledge to get from that point A to that point B."

I nodded, but I wondered how far from point A this kid had really come. He was still a grad student in Cambridge playing with stuffed monkeys for his grant money. Semyon, on the other hand, had indeed made millions. But he'd also taken risks, had paid a price that maybe this kid wouldn't understand.

"Radar was invented just a few buildings over," Evan continued, his face beaming with what could only be described as techno-pride. "And many more things came out of this place that you'd need clearance to uncover and even more clearance to talk about. Hell, you can't fire a nuke at a third-world country without some PhD student from MIT checking over your trajectories."

I wasn't sure it was something to be proud of, but I understood his point. I had come to the Media Lab to get a better understanding of what it was that made Semyon uniquely qualified to do what he had done. It wasn't just that he was smart, that he knew math and computers like no one else. He was also a hacker at his core. Not only did he know math, he knew that math could make you rich.

Along the way, however, he learned something else. Something that Evan, for all his pride, didn't understand. Something that maybe the guys who made radar and built guidance systems for ICBMs learned much as Semyon did, step-by-step.

Sometimes, in some circumstances, math can also get you killed.

CHAPTER 5

Ten Thousand Feet over Nevada

The moment came suddenly, so suddenly that Semyon half thought he was still dreaming. His face pressed against the cold, glass, egg-shaped window, his long legs pretzeled up against the seat in front of him, his arms tucked beneath the thick folds of the oversize puffy jacket he was still wearing—even though it was deep summer and he was jammed in the second-to-last row of a sweaty, overheated America West 737—and his face mostly hidden beneath a gray flannel cap. His eyes were open, but his vision far away, lost in the darkness outside the window. And then, as the plane banked hard over the pitch-black desert, suddenly there it was, the moment, singular and defining. A flash of bright white light hit his vision, and then the desert seemed split in half. An electric, Technicolored scar appeared, a twisting river of neon snaking off across the blackness, so bright it made Semyon's eyes water.

"Holy shit," he mumbled, blinking.

Vegas. The place he'd been dreaming about ever since his indoctrination into Victor's little cabal. The airplane dipped lower, sending a lurching feeling through Semyon's stomach, but he barely even noticed. He was minutes away from beginning the journey that had been consuming his life for the past month. Five hours ago, he had boarded the 737 in Boston, but even then, it had seemed so unreal.

Now it was finally hitting him. This was not another test, not a checkout or a lesson in some MIT classroom. This was the real thing.

Beneath the neon, some of the larger hotels came into view. Even from that distance, thousands of feet in the air, Semyon was awed by the scale of what he was seeing. He could make out the replicated Roman architecture of Caesars Palace, the boxy glow of the Flamingo Hilton, the teetering spire of the Stratosphere, the elegant glow of the Desert Inn. He could see the headlights from the cars lined up along the strip, the classic snarl of traffic that was more and more becoming the city's nightly routine. So many people on their way to the casinos, so many tourists, their wallets bulging with cash, all of them hoping for a big score, praying that the cards and the wheels and the spinning slots went their way. Just this once, just this time, just this night.

They might as well have dumped their wallets on a bonfire and watched them burn. Semyon knew that the tourists were there only to fuel the electric scar. Victor called them "civilians," but Semyon could tell by his tone what he meant by the term: *losers*. Because, inevitably, that's what they were. They came to the neon city to lose.

Semyon turned away from the window, stretching his neck as he tried to untangle his legs. The motion sent his knee inadvertently jerking out to his left, hitting the extended plastic tray table of the seat next to him. The sound echoed through the quiet cabin, turning heads in most of the nearby rows.

"We're not exactly built for coach-class seating, are we?"

Semyon smiled sheepishly as he watched Allie stretch awake next to him, her own long legs contorted beneath the tray table he had nearly just rattled. She ran a hand through her short hair, then blinked at the window beyond his shoulder.

"Looks like we're almost there. Even though I slept through most of it, this flight's about an hour too long."

Semyon nodded. Though in his mind, the flight had gone pretty fast. The first two hours he and Allie had dealt cards to each other, going over and over again the three techniques that Victor had taught them during the past month. The next hour had been spent in quiet conversation, working on their "characters"—practicing their fake names, inventing fake backstories, making up little facts they could throw into conversation to make their story seem more true. Semyon wasn't sure why Victor had sent him and the blonde beauty on ahead, but Victor and Jake wouldn't arrive in Vegas for at least three hours after Semyon and Allie's flight landed. That meant Semyon and Allie would have a chance to try out what they had learned in a real casino, with real money—on their own. They would be going in as a team, working together to beat the blackjack tables the way Victor had taught them. Playing the roles that Victor had invented.

Semyon made his voice as deep a baritone as possible. "Although it wasn't as bad as the flight from Moscow, was it, my dear?"

Allie put on her best Russian accent. "Of course you're right, Nikolai."

They both laughed as the captain's voice came over the intercom, explaining that they had begun their final descent into McCarren. It was pretty crazy, what they were about to do. As soon as they landed in Vegas, they weren't going to be Semyon Dukach and Allie Simpson any more; instead, they would be Nikolai Nogov and Sonya Fedorov, on their first visit to the U.S. from Mother Russia. They'd no longer be MIT students earning math and computer degrees. Instead, Semyon would be playing the part of a young Moscow "businessman," whatever that might mean. And although Allie didn't know any Russian, Semyon was certain she would have no problem passing as a Russian party girl.

Still, even though they could probably look the parts that Victor had drawn up for them, Semyon wondered if they'd really be able to pull the role-play off. He wasn't concerned about the math behind the three techniques—heck, he and Allie had practiced so much, the numbers were second nature to him. But the character acting, that was something he'd never really tried before—and to pull this off, his character would have to be perfect. If he let his character slip, well, he didn't know what would happen, but he was certain he didn't want to find out.

He pushed the button on his armrest, moving his seat back to its upright position. Allie followed suit, then leaned over him to get a better view out the window. Her perfume hit Semyon full on, and he felt his cheeks getting warm. He knew the perfume was part of her costume, like the thick red lipstick she had carefully applied right before takeoff, and the short leather skirt he was scrupulously trying to ignore; but even so, it sent thrills down his spine.

"So what do you see when you look down there?" she asked, gesturing toward the lights, which were growing in intensity with each dip in altitude.

Semyon wasn't sure what she meant. What did he see? He saw Vegas. A city full of casinos, of strip clubs and discos and restaurants, the ultimate party town, an embodiment of every vice ever conceived. More than that, he saw the blackjack tables, the cards flipping and fluttering through the air, a moving tapestry of aces and tens and kings and queens pirouetting against a green felt canvas.

"I see money," he responded, shrugging. Around him, he could hear the flight attendants readying the cabin for landing, and beyond that, the eager chatter of two hundred other coach-class passengers coming to life. It was a palpable thing, those minutes before landing in Vegas, when all the stored energy of those five hours started to bubble up. Everyone was ready to get out of that airplane and onto that strip. "I see the money we're gonna make."

"Well, yeah," Allie said, pushing her tray table closed and sitting up in her seat. "But I think what we're about to do, it's a lot bigger than that. We're going to take on the casinos. We're going after the very system of it, you know? What we're doing, it's supposed to be impossible. Beating the system. And the subterfuge of it, you know, the secrecy. The fake IDs, the costumes—fuck, wearing hooker makeup, a short skirt, stripper heels. And the idea, well, that we're the only two people on this plane who actually have an advantage . . ."

She laced her fingers together over her bare knees. Her face was flushed with the effort of trying to explain her thoughts. Then she shrugged.

"Well, fuck, yeah, the money, too. I've got nothing against the money."

Semyon laughed. The plane was descending fast now, they were just a few minutes from the runway. Allie looked at him, her expression more serious than before.

"I'm on full scholarship, you know. My father is a math teacher at an elementary school in Somerville. I've never really had much money, nor have I really missed it. But it would be nice, for once, to know what it's like to not care what things cost. To be rich."

Semyon raised his eyebrows. Her admission had taken him entirely by surprise. Though they had been practicing together for four weeks, they had barely spoken about their pasts. He still knew Owen way better than his teammate, still saw Allie as a barely approachable mystery. This was the first time she had even come close to opening up to him, and it took him a full minute to respond.

"I kind of thought—"

"You thought I was some rich kid from some preppy suburb," she said, laughing, "driving to school every day in a bright red convertible with my Harvard boyfriend in the passenger seat. Yes, I know, I give off that appearance."

Semyon was really beginning to like her. Maybe he had been wrong to see her as unapproachable. Maybe they had more in common than he'd realized. In any event, he felt more comfortable with her than at any time before. In fact, he hardly even noticed that her skirt had now ridden halfway up her alabaster thighs.

"Actually," she said, still smiling, "I did have a Harvard boyfriend. He was boring as shit, but he did have a nice car. I got rid of him when Victor told me I'd be able to afford my own car within a few weeks of joining his team."

The landing gear touched down without warning, and Semyon grabbed at his armrests. Allie reached out and put a comforting hand on his, then gave him a playful wink.

"Appearances can be deceiving, eh, Nikolai? Isn't that what this is all about?"

Semyon's hands were shaking as he slowly unzipped his pants. He could feel the sweat running down his back as he struggled to get the too-tight jeans down around his hips. The thick material somehow caught on the elastic of his boxer shorts and then, finally, the jeans came loose, and he pulled them down to reveal the twin black Velcro pouches that were wrapped tightly around his upper thighs.

He took a deep breath, leaning back against the cold metal wall. He was in a cramped bathroom stall on the first floor of the airport terminal. His duffel bag and bulky winter jacket were both hanging from the hook on the stall door, leaving him very little room to maneuver. Every time he turned, he banged his ankles into the porcelain toilet, causing him to unleash a tirade of curses at the architects who'd made these fucking stalls so fucking small. At least he remained in character, the curses coming out in filthy Russian he hadn't used since junior high.

He calmed himself as he wriggled a bit to get his jeans lower down his legs. He'd have been much happier wearing baggier pants, but Victor had assured him that the Velcro packs were actually better concealed beneath tighter denim. And Victor hadn't steered him wrong; the security agent at Logan Airport had barely glanced at him as he'd gone through the metal detector, and thankfully hadn't even patted him down.

As the jeans slid toward the stall floor, Semyon reached for the Velcro pack on his right thigh and carefully undid the flap. The tearing sound echoed loudly in his ears, but he was pretty sure the other travelers in the bathroom wouldn't think anything of it. It was an unwritten rule of public restrooms; people usually ignored the sounds that came out of other people's stalls.

He reached inside the pack and felt the stacks of bills with his fingers. A warm sensation moved through his muscles as he carefully lifted the first stack free. He took another breath, his nostrils filling with the distinct smell of cash. It was interesting, that money had a smell, and he wondered if there was a way to tell denominations simply by scent. He wondered if this stack of hundred-dollar bills somehow smelled different from any stack of bills he had ever held before because it was worth so much more.

Ten thousand dollars, one hundred hundred-dollar bills banded together by a strip of bright blue paper. *Ten thousand dollars.*

Semyon shivered as he reached forward and jammed the stack of bills into the exposed outer pocket of his jacket. Then he reached back into the Velcro pack and grabbed another stack.

In just a few short minutes, he had removed the contents of his packs and transferred the cash to the pockets of his bulky jacket. Then he carefully snapped the pockets shut and put the jacket back on, slinging the duffel over his shoulder. The duffel was pretty much worthless, containing a few changes of clothes, a notebook, and a handful of false IDs. But the jacket was now the most valuable thing he'd ever carried. It felt twice as bulky as before, bulging even more from the stacks of hundred-dollar bills. Twenty of them, to be precise. Two hundred thousand dollars. About half of it was Victor's own stash, blackjack money he had earned refining the three techniques. The other half had been raised from investors who were looking for more of a return than the standard card-counting team would give them.

Two hundred thousand dollars. And that was only half of the stake he and Allie had brought with them to Vegas. Allie had two hundred thousand more, though Semyon wasn't sure how she had hidden the money, considering she was wearing such a tight leather skirt. He hadn't seen her without her jean jacket, but he was pretty sure she had some sort of tank top underneath. Then again, he was pretty sure airport security spent less time working over girls like Allie; she might very well have had the money banded around her waist.

He zipped his jacket halfway up his chest, then prepared to head back out into the airport. He wondered if Allie was already on her way to the meeting place. On Victor's suggestion, they had decided to split up the minute they left the plane. Victor had assured them that there was no reason for them to expect any sort of surveillance, but it was always good practice to remain cautious. The two of them had no reason to be seen together coming off a plane from Boston, or to be together in the airport; their act began at the casino, not in a domestic airport terminal. And again, their act had to be perfect. Because even though there wasn't any surveillance waiting for them at McCarren, people would be watching. Casino hosts, pit bosses, security personnel in the booths upstairs, eyes in the sky. Not to mention the private eyes who worked for the casinos, whose job it was to protect the money from schemers like Victor and his lot. Semyon knew all about the Griffin Agency, the most notorious PI firm in Vegas, which worked, in some capacity, for nearly every casino in the world. Semyon knew all about the cat-and-mouse game he would be playing with the people on the other side. He knew that in Vegas, someone was always watching.

The truth was, that thrilled him way more than it frightened. *Let them watch.*

Semyon set his jaw and yanked open the door to the stall. Then he straightened his two-hundred-thousand-dollar jacket over his shoulders and headed on his way.

Let them fucking watch.

CHAPTER 6

The Mirage
Las Vegas, Nevada

he heat took him completely by surprise. Searing, arid, a blast so sudden and vicious it knocked Semyon back on his heels, nearly sending him colliding into a Japanese family that had surreptitiously taken up position right behind him and were now furiously taking photos. Then, a second later, came the flames, and Semyon instantly forgot about the heat. Towering into the sky, huge bright orange ropes of fire twisting and turning, erupting higher and higher from the mouth of the huge volcano. A gasp erupted from the crowd of people, mingling with the constant click of the Japanese family's cameras and the steady whir of a half-dozen oversize camcorders.

Semyon stared at the flaming volcano, shaking his head. He had read about the thing in a guidebook a few days before boarding the flight to Vegas, but the writer had certainly failed to capture the sheer enormity of the construct. The volcano was fake-looking, to be sure, but it was still pretty damn magnificent. Maybe twenty, thirty feet tall, at least fifty feet in diameter, and surrounded by a small moat and a waist-high fence. And the flames that spouted from its cavernous maw, well, if they were fake, they certainly didn't feel like it. The temperature in the air had gone up a good twenty degrees with the first blast—and since it was June and the middle of a fucking desert, that meant it had to be somewhere near a hundred and ten, easy.

What was worse, Semyon was still wearing the bulky winter coat, which he didn't dare take off. And he'd been circling the monstrous volcano for more than ten minutes, winding his way through the crowd that had slowly gathered as the eruption time had neared. Circling and circling, sweating and sweating, as he searched for Allie. In retrospect, the volcano was not the great meeting place that he and Allie had assumed. The sheer size of the monstrosity, not to mention the crowd, made it a little like looking for a needle in a haystack. A lanky, sexy blond needle in a haystack full of blistering-hot magma.

Semyon pushed past the Japanese photog family and started another circuit around the volcano. He felt minorly light-headed as he struggled through the crowd of tourists, which was now pressed even tighter together, watching in awe as the flames spouting upward began to change colors. "Oohs" and "aahs" preceded each shift, from orange to red to green to blue. Semyon didn't have to look at the flames to see the colors, he could see the glow reflected in the wide-eyed faces that were blocking his way. The civilians were certainly getting their money's worth, here outside the Mirage. If they never went inside the casino's doors, they would go home ahead. But Semyon knew that wasn't going to happen. Nobody came to Vegas to stand outside on the sidewalk, no matter how amazing the view. Especially during the summer.

As for himself, he couldn't wait to find Allie and get the hell into someplace air-conditioned. If he didn't find her soon, he was pretty sure he was going to die from the heat—

"Nice night for a stroll, eh, Nikolai?"

And there she was, at the edge of the crowd, waving at him from between two overweight tourists in matching bright green shorts and sloppy white T-shirts. She had shed her jacket, which was draped over her extended left arm, and with her long, bare legs and skimpy tank top, she was quite a sight. More than a few eyes turned away from the volcano to watch her work her way to where Semyon was standing. Then, when they saw the crazed, overcooked Russian kid in the bulky winter coat, most of them quickly looked away.

"You might want to invest in a lighter jacket," Allie said as she grabbed his hand and led him away from the volcano toward the Mirage's front door. "Although the pre-heatstroke glow is starting to grow on me."

Semyon was immediately conscious of her fingers against his palm, the way she had immediately taken charge and crossed the boundary of two near strangers. Of course, to pull this off, they had to appear like a couple, of sorts, from the minute they walked into the Mirage. But the ease with which she had broken down that wall, the ease with which she had initiated contact—it was pretty damn thrilling. He was liking this blackjack thing more and more.

A moment later, they moved through the glass threshold of the casino and entered the bustling lobby. A blast of frozen air hit Semyon full in the face, taking his breath away. The contrast of the hot, arid air outside and the overly air-conditioned interior of the casino was mind-numbing, and it took Semyon a full second to regain his senses and survey their surroundings.

He knew what to expect from the Vegas guidebook, but still, the Mirage was impressive. Ahead of him, a faux forest spread out across the huge, crowded atrium. Trees, bushes, grass, bisected by little stone pathways, complete with little bridges and bubbling brooks. He could hear a waterfall somewhere in the distance, and from somewhere up above, the chirping of birds. Canned birds, to be sure, audio pumped through speakers hidden somewhere in the curved, faux sky that took the place of the casino's ceiling, but pretty convincing nonetheless.

Still, Semyon hadn't come here for the birds. Beyond the forest, he could make out the real heart of the "mirage": the casino floor. Row after row of slot machines, gleaming in the fake daylight. And behind the slots, the table games, stretching for what seemed like hundreds of yards in every direction. Roulette, craps, Caribbean poker, three-card, and of course, blackjack. The tables looked crowded, well more than the row after row of slots. The Mirage was a good place for table play, one of the most popular megahotels on the strip for people with money to burn. Though most of the crowd seemed downscale, there seemed to be a fair number of better-dressed players. It was a good place for a pair of high rollers to start.

Semyon looked at Allie, then grinned. His heart was pounding in his chest.

"Let's do this," he said, because he couldn't think of anything more dramatic.

Numbers were his thing, not words.

She nodded, and they moved quickly through the forest, winding past the frat boys in sweatshirts, the cocktail waitresses in tiny black skirts and brightly colored blouses, the Midwestern couples in matching Disney T-shirts, the bachelor parties on their way to being way too drunk, the bachelorette parties already there. Past the men in suits, the women in dresses, the young and the old, the Friday-night Vegas crowd. Past the rows of slot machines, past the craps tables and the roulette wheels. Past the rows of low-stake blackjack tables, the quarter, nickels, and dimes, all the way to the back of the casino—to a roped-off area with a curved, faux-wood entrance.

The high-stakes lounge. Semyon paused before they entered, unzipping his jacket, revealing his unbuttoned white tuxedo shirt, which hung sloppily over his tight jeans. The shirt he had borrowed from Owen, the jeans he had bought at a used clothing store on Newbury Street. He knew the look was a bit disjointed, but that, too, fit the part. He knew a little bit about Russian high-roller types; growing up in New Jersey, he had taken a few trips to the New York City borough of Queens, to the Russian community known as Little Odessa. He had seen many Russian "businessmen" drinking in the small Russian cafés that lined the streets of that part of the city. He remembered how they'd seemed always to dress in pieces; all the parts of their outfits expensive, but none of them a set, nothing that really matched. Expensive designer shirts with jeans, Bulgari shoes with white gym socks, tuxedo pants with wife-beater tank tops. It was the fashion of men who had gotten rich so fast, their sense of culture and style never had a chance to catch up. The new breed of Moscow millionaires—that was Nikolai Nogov, that was the character Semyon wanted to create. Not only in the clothes, but in every way, he had to become this man, this high roller. He had to forget everything about him that screamed MIT.

He glanced at Allie, at the way she pushed out her chest, her chin held high, her bright red lips pursed above her bright white teeth. With her on his arm, he had a feeling it wasn't going to be all that hard to distance himself from MIT.

He crooked his elbow around hers and strolled into the high-stakes lounge.

CHAPTER 7

The Mirage
Las Vegas, Nevada

ike banded stacks of hundred-dollar bills, the high-stakes lounge had a distinct smell; actually, a combination of scents, familiar to anyone who had spent time with people of a certain economic status—people with money to burn. First, there were the twin olfactory standards: cigar smoke and booze. The former, the thickly sweet pall of tobacco that had been grown within a few dozen miles of Havana, the latter just as thick and sweet, primarily Scotch, single malt from some grassy highland nobody had ever heard of. On top of that, a liberal whiff of perfume, certainly French, utterly expensive. There was also cologne, but the type of men who wore cologne usually worked for the type of men who frequented the high-stakes lounge. More compelling, more defining than the perfume or cologne, there was the underlying, overwhelming scent of leather; shoes, belts, wallets, the chemical memories of couches and car seats. Even more than the tobacco or the Scotch, it was the leather that linked the high rollers together; leather was the true symbol of modern money.

Semyon Dukach had never owned anything that was made of leather. As he moved across the threshold of the high-stakes lounge, he hoped that Allie's little skirt would be more than enough for the two of them.

The lounge was circular, the walls and ceilings done up in wood tones, with plush curtains and brass and bronze fixtures. There was a curved mahogany bar along one side, staffed by a handful of cocktail waitresses in very short skirts. The lounge was much less crowded than the rest of the casino, though there was still a fair amount of gambling going on. Semyon counted a dozen blackjack tables running in a tight semicircle around the front of the room; he assumed the other high-stakes games had their own areas somewhere else in the casino. None of the blackjack felts was empty, but a few had only single players. The casino staff seemed somewhat skeletal compared to outside; aside from the dealers and cocktail waitresses, there were only two pit bosses in the center of the room, keeping track of the action. One of the men was tall, maybe sixty years old, with thick, snow-white hair and a wrinkled face. The other was shorter and stocky, balding, with a noticeable gut. Both were wearing dark suits with even darker ties. The one with white hair had a phone up to his ear, while the fat one strolled back and forth behind the dealers. Neither took much notice of Semyon and Allie as they entered the lounge—but Semyon was pretty sure that would soon change.

He took Allie by the hand and led her in a slow stroll behind the blackjack felts. He pretended to watch the play, the flow of the cards, the click of chips being thrown into the betting circles, but really he was concentrating on the dealers. Because for what he and Allie were about to attempt, it was really all about the dealers. As Victor always said, the cards are predictable; it's the dealers that keep changing.

There was a pretty wide variety in this room; about half men, half women, of varying ages and ethnicities. Most of the dealers looked to be in their mid- to late thirties, a few were older, and one woman could not have been much past twenty-two. The women were all fairly attractive; two were blond and actually quite pretty. The men were clean-cut, though a few sported jewelry ranging from pinkie rings to gaudy bracelets and necklaces.

Semyon had gone halfway around the semicircle of blackjack tables when he felt a tug on his wrist. Allie was aiming her chin at one of the felts, farthest from the entrance to the lounge. Her blue eyes had narrowed, and her breathing had gone soft.

"What do you think?"

Semyon glanced at the table. There was one player seated in the third-base seat at the table—farthest from the shoe—a thickset man, with thinning black hair. He was hunched forward over the table, his gray suit bunching up around his oversize paunch. He looked to be middle-aged, probably in town for a convention. He wasn't wearing a wedding ring, but in Vegas, that didn't really mean anything. Especially among the conventioneers.

Semyon moved his eyes past the portly player to the dealer. The dealer was a woman; short, thin, delicate features, with frosted blond hair and too much eye shadow. She was wearing a name tag that identified her as Fran, stating that she was a native of Las Vegas. Fran couldn't have been more than five foot two, a hundred pounds at most. She was fast with the cards, her arm moving in a smooth, controlled arc. When she beat the player, she swept up his chips at near warp speed. She was skilled—but that didn't make any difference. What mattered were her hands. Her *small* hands.

Semyon smiled, understanding why Allie had singled out that particular table. Allie wanted them to try Victor's first technique.

Semyon squared his shoulders and started toward the table. The first technique was as good as any place to start, and the miniature dealer seemed perfect for the scheme. They would need the table all to themselves, but Semyon wasn't worried about the portly conventioneer.

When Semyon got within a few feet of the table, he let go of Allie and reached into his jacket pockets with both hands. As he dropped onto the stool closest to the shoe—the first-base seat—he began tossing the stacks onto the felt. Fran was midway through dealing to herself, about to turn over a card to add to the sixteen she had showing. She froze middeal as she watched the growing pile of hundred-dollar bills. Within a few seconds, Semyon had one hundred and fifty thousand dollars on the table. Fran blinked, then finished out her hand. She paid off the conventioneer, who barely even noticed that he'd just won a hundred bucks. Like the dealer, he was staring at the pile in front of Semyon.

Semyon pushed the pile of wrapped bills forward across the felt, at the same time grabbing Allie by the waist and helping her onto the stool next to him. Then he slapped a hand against her bare thigh and shouted in a loud, heavily accented voice: "We're going to have some fun here, yes?"

His ears were ringing from his own voice—and the feel of her skin against his palm—but he did his best to keep his composure. He watched as Fran turned to look at the two pit bosses. Both of them were staring at Semyon and Allie. Actually, most of the high-stakes lounge was looking in their direction. Even the few truly wealthy patrons in the room were impressed by the pile of cash on the blackjack table.

Suddenly the two pit bosses snapped into action. The man with the white hair hung up the phone and scurried toward Semyon's table. The bald pit boss headed straight for the bar, corralling one of the prettier cocktail waitresses. Semyon smiled inwardly. It was going just as Victor had said it would. This wasn't some card-counting scheme that relied on secrecy, camouflage, flying under the radar. This was a nuclear attack. Semyon didn't want to fly under the radar. To pull this off, he wanted to be as big a fucking blip as he could possibly be.

The pit boss sidled up next to the dealer and smiled widely as Fran began counting the stacks of hundred-dollar bills.

"Welcome to the Mirage. I don't believe we've had the pleasure of your company before?"

Semyon shrugged as he watched Fran turning the cash into a rainbow of brightly colored chips. Orange thousand-dollar denominations, purple five-hundreds, black hundreds, and even a few multicolored squares that represented five thousand dollars each. The felt quickly began to look more like a Jackson Pollock painting than a blackjack table.

"We are just in from Moscow," Semyon said, taking the stacks of chips from the dealer. He piled half up in front of him, half in front of Allie. "This is our first time in Las Vegas. It is very nice. We like it very much."

The white-haired pit boss showed more teeth as he watched Semyon push a thousand-dollar chip into the betting circle in front of him, then another thousand into the circle in front of Allie.

"Well, I'm sure you'll be very happy here at the Mirage. And of course, we'd love to set you up in one of our VIP suites for the weekend, Mr.—"

"Nogov. You can call me Nikolai. This is my friend Sonya."

Semyon tossed the man one of the ID cards Victor had given him, stifling the sudden urge to crack up at his own theatrics. He was channeling Boris Karloff, with a little James Bond thrown in for good measure. He half wanted to ask for a Bellini, shaken, not stirred. He reminded himself that this was more than a game. He had just exchanged one hundred and fifty thousand dollars for casino chips. Enough money to set his family up for two years. Enough money to pay back all his student loans, in one fell swoop. No, this wasn't simply a game. He watched the pit boss as the man looked over the ID; it wasn't a driver's license or a passport, but the man didn't seem to care. One hundred and fifty thousand dollars bought Semyon all the authority he needed.

Semyon glanced at the conventioneer at the other end of the felt, who was momentarily preoccupied by Allie's bare legs. Then Semyon leaned closer to the white-haired pit boss, lowering his voice.

"Excuse me, but might it be possible to, um, raise the minimum bet a bit? Perhaps five hundred a hand?"

The pit boss raised his bushy white eyebrows. Then he looked at the conventioneer. He understood. It wasn't really about the minimum bet, it was about clearing away the riffraff. It was a move Victor had taught Semyon during his training. Often high rollers used the request to clear the table of unwanted gawkers. The higher the minimum bet, the less likely it was that other players would interrupt a high roller's fun.

The pit boss took two steps to his right and pulled a plastic card out from behind the table. He placed the card onto the felt, then tapped at it with one manicured finger until the conventioneer looked up from Allie's legs and read the embossed script: *Minimum, 500—Maximum, 10,000.*

The conventioneer was about to protest, when the white-haired pit boss gently patted his shoulder.

"Mr. Cox, I'm sure you'll be much happier at one of our one-hundred-dollar tables. But if you'd rather take a break, I'd love to comp you dinner at our new steak house, and follow that up with a front-row seat at our showgirl revue. Would that be agreeable?"

The conventioneer glanced at Semyon and Allie, then shrugged, scooping up his chips. Semyon smiled inwardly as the pit boss escorted the man away from the table. Semyon was running the show. He winked at Allie as Fran rapidly began dealing out the next hand.

The cards came quickly, and Semyon barely paid attention to the numbers as he and Allie played their two hands. After so much practice, the decisions were rote, like a computer spitting out predecided commands. It was called basic strategy, the optimum moves a player could make to enhance his odds at the game of blackjack. When followed exactly—and Victor's protégés had to master perfect basic strategy to even be considered for further training—BS left the player at a very small disadvantage to the casino, around 2 percent per hand. That meant that at a one-thousand-dollar minimum bet, Semyon and Allie could expect to lose around twenty dollars per hand. The variation was high, so at any given time, they could be up or down much more than that. But if they played long enough, their losses would near twenty dollars per hand. With a fast dealer, this 2 percent loss rate could add up pretty quickly. They could expect to lose about two grand per hour.

Card counters could reverse the loss rate by keeping track of the high cards. A card-counting team could gain about 2 percent on the house; Victor's larger team of MIT kids would earn about what Semyon could expect to lose—twenty dollars per hand at a thousand-dollar bet.

But Semyon and Allie weren't counting, though certainly they knew how. They weren't interested in a 2 percent advantage. They were looking for something much bigger.

The cards kept coming. When the numbers called for a hit, Semyon did so with fervor, hitting the felt hard enough to make the players at nearby tables look up. When he won, he shouted in Russian, clapped his hands, made an emotional show. When he lost, he cursed, slapping the table with his hands. Sometimes Allie squealed at his antics, playing the part of the passionate party girl. Somewhere near the end of the shoe, the cocktail waitress arrived with a tray of drinks, and Semyon took two for himself, two for Allie. By the time the shoe finished and Fran began the shuffle, Semyon was pretty sure the pit bosses had already written him off as crazy, drunk, and rich. The perfect cover for what he was about to do.

He watched as Fran's small hands worked at the cards, stacking them in a high tower, then splitting them off into two equal piles. She began riffing them together, effectively mixing them up. It was an interesting thing to watch—especially considering that Semyon had been practicing that exact same shuffle six hours a day for the past four weeks. In Vegas, a blackjack dealer's shuffle was as routine and choreographed as a Russian ballet. A few casinos had their own slight variations—but dealers, as a rule, used a ritualized method to shuffle the cards. Victor had studied that shuffle for years and had passed what he had learned on to his protégés. That knowledge, along with the application of some impressive math, was the basis for the three techniques.

Fran finished riffing the cards together and carefully rolled the long pile of cards into a horizontal stack that stretched about a foot across the green felt. Six decks stacked together, effectively mixed up from the past deal, ready for a new round. Semyon was sitting close to her in the first-base seat, and his attention subtly moved to her left hand, where she held the back of the rolled-over shoe. She was barely covering the bottom card with her little palm. With little effort, Semyon could see a little sliver of red peeking out from between two of her fingers and the curve of a number. A three. The bottom card was a three of hearts. Not a card that mattered, not a card that he could use. But still, the fact that he could see it was staggeringly important. Because sooner or later, that card would be something he could use. And that would make all the difference in the world.

Fran took out a plastic yellow cut card from her chip tray and handed it to Allie. Allie smiled, cutting the deck about two inches from the back of the horizontal stack. It looked like a random placement of the card, but the truth was, it was anything but random. Allie—and Semyon—had been practicing cutting decks almost as many hours as they'd practiced shuffling cards. Both she and Semyon could cut a deck of cards to an exact spot nineteen times out of twenty. If she wanted to, she could cut to the fifty-third card nearly twenty times in a row. By the looks of things, she had cut to around sixty cards. Fran split the deck where Allie had cut it and moved the front end of the deck to the back.

The three of hearts was now sitting next to the sixtieth card in the deck. In other words, both Semyon and Allie knew that the sixty-first card in the deck was a three. Not an important card, not information they were likely to use. But that didn't matter, at the moment. The three techniques were opportunistic. The opportunity hadn't yet arrived, but the techniques were already in motion.

Fran placed the cards back in the shoe and began dealing. Semyon and Allie went through their act again, playing their thousand-dollar bets, winning and losing and winning and losing. The cards kept coming, and Semyon kept playing his role, shouting when he won, cursing when he lost. At one point, after hitting a blackjack, Allie actually jumped off her stool and grabbed Semyon in a huge hug, knocking both their drinks off the table and onto the floor. The pit bosses scurried over to make sure the cocktail waitress was there to replace the drinks before the next hand.

The second shoe finished with a few good hands, and as Fran began her second shuffle, Semyon took quick stock of how they'd done. They were down a little, but nothing spectacular; he had lost two thousand dollars, and Allie had won fifteen hundred. Altogether, they had lost five hundred dollars in twenty minutes of play.

Semyon looked up from the chips in time to watch Fran finish the last riff and roll the cards into the long stack against the felt. This time her hand was covering the last card more effectively, and he couldn't see color or shape. But he wasn't discouraged. He was opportunistic, but he also had a few tricks up his sleeve to help the opportunity along.

He waited until Fran picked up the plastic cut card. This time she held it out toward him, expecting him to take it and make the cut. He quickly held up his hand, palm out.

"No, no, no. Lady brings the luck."

Fran smiled, shifting the cut card toward Allie. To her surprise,

Allie quickly shook her head. "No, Nikolai, I am not feeling lucky now. You take it."

Fran moved back toward Semyon, but he crossed his arms against his chest, an angry look on his face. He snapped out in Russian, something about "Sonya" being an ungrateful bitch. Allie didn't respond—well, she didn't exactly understand what he had said, but she certainly got the gist—but instead, stood up and suddenly slapped the back of her hand across his cheek.

He felt his cheek turning bright red where she'd hit him, and for a brief moment he couldn't find his breath. He hadn't quite expected that. They'd rehearsed the moment before, in a classroom at MIT, but she'd never actually hit him before.

If Semyon was shocked by her sudden move, Fran, the dealer, was completely stunned. The cut card was frozen midair between Semyon and Allie. Fran's eyes were wide open, and it was obvious she had no idea what to do. She glanced back toward the two pit bosses, and in that split second, Semyon dropped his gaze to her other hand, where it held the back of the shuffled stack of cards. Just as he'd suspected, her fingers had inadvertently moved apart, and he could now easily see the back card. His stomach tightened as his heart began to pound.

An ace of hearts.

It couldn't have been more perfect. The ace was the most powerful card in the deck. Semyon quickly composed himself and, with a swift motion, grabbed the plastic cut card out of Fran's hand.

"I guess I will cut, since my lady friend seems to be in a bit of a mood."

Fran turned back to Semyon, obviously relieved that the argument had been resolved. Allie quietly sat back on her stool, watching as Semyon swiftly cut two inches out of the deck. In his head, Semyon could see the cards separating at exactly the fifty-ninth card, just as he'd practiced. He watched as Fran moved the deck to the back. This time she burned one card, tossing the top card into the discard pile. Then she rolled the deck into the shoe.

Semyon took a deep breath. He now knew, with near absolute certainty, that the fifty-ninth card in the deck was an ace. It was just a matter of taking advantage of that knowledge.

He rubbed his hands in his hair, then slapped both palms against the table.

"You know what? I'm feeling like really playing now. No more fooling around."

He took three five-hundred-dollar chips and placed one in each of the three closest betting circles. He watched as Allie mimicked his move, taking up the next three betting circles with her own chips. Now they were playing six hands at once, nearly the entire table. The dealer looked back at the pit bosses, who both nodded; they were going to let Semyon play as many hands as he wanted. He was the dream player, an emotional high roller with a seemingly endless bankroll and a wild style of play. Surely, he'd be pouring plenty of money into the casino's coffers.

Fran dealt out the first hand, and Semyon casually watched the cards, trying not to appear to be paying too close attention. The truth was, he didn't really care what the cards were—just the number of them that were going by. He played out his hands, hitting when he was supposed to, staying on the good hands, and Allie did the same. After the first round, twenty-one cards had gone by, including the dealer's. Fran dealt again, and after two rounds, the number of cards that had passed by was forty. Again, Semyon and Allie put out their five-hundred-dollar bets, and six more hands were dealt. This time around, Semyon made sure not to hit any of his hands—even though one of them was just a twelve, and the dealer was showing a nine. Allie did the same. Her play seemed even more bizarre, since one of her hands added up to seven—and a hit couldn't possibly have hurt her. But she made some comment about trusting the fates, and Fran didn't seem to care. The truth was, dealers saw so many pathetically bad players, they were more surprised when players got it right than when they got it wrong.

After the third round, fifty-five cards had come out. That meant that if Semyon had cut correctly, the fifth card that came out of the deck was going to be an ace. Since he was playing the first three hands, that meant the ace should land on Allie's second hand. The reason he and Allie were playing the whole table was to ensure themselves in case his cut was a little off. If he was a card off in either direction, either he or Allie would still get the ace.

Semyon leaned back on his stool. Now was the time to make his move. Nothing subtle, nothing secret. In full view of the cameras that lined the ceilings, the pit bosses, the eyes in the sky. Semyon grabbed at his stacks of colored chips and suddenly pushed three stacks of ten thousand dollars into the betting circles. Without a word, Allie followed suit.

Fran stared at their bets, then once again looked back at the pit bosses. Both pit bosses nodded at the same time. They had been watching Semyon's play all along. Now the crazy Russian and his equally crazy girlfriend had put sixty thousand dollars on the table. It looked like it was about to be Christmas for the Mirage.

Fran reached for the cards and began dealing. Semyon's first card was a king, his second card a seven, his third card a ten. Allie's first card was a four—and then bang, there it was, the ace of hearts. Allie's third card was a king. But Semyon didn't really care about any of the cards because he had done his job properly. He had cut to an ace and had run the table to get it dealt to his partner. Nothing else really mattered because it was really all a matter of math.

An ace dealt to a player's hand gave a 51 percent advantage to that player. The other five hands at the table were playing at the normal 2 percent disadvantage—so in total, Semyon and Allie had sixty thousand dollars bet at a mathematical advantage of 45 percent. Not 2 percent, like card counters, or negative 2 percent, like basic-strategy players. *Forty-five percent.* It was a huge number. It had taken them about half an hour of play to find an opportunity like this—a perfect moment to use the first technique. But there was no denying the power of what Victor had rather unceremoniously dubbed "cutting to the ace."

Fran continued to deal the cards, giving herself a seven, then dutifully doling out the rest of the hands. Semyon managed to pull two twenties and a seventeen, all of which he would hold. Allie pulled a nine on her first hand, for a thirteen. Then out came a king—giving her a blackjack on the ace Semyon had guided to her. She clapped her hands, squealing loud enough to let Semyon know her pleasure wasn't an act. Her last hand added up to fifteen. After the dealer got her card, Allie hit on her first hand, busting, and hit on her third hand, drawing a three for an eighteen. The dealer turned over a ten, for seventeen. Semyon won all three of his hands. Allie won two, the blackjack paying fifteen thousand. Altogether, in that single hand, Semyon and Allie won forty-five thousand dollars. More than their expected win, but again, the numbers didn't matter. The math was certain. The first technique gave them an incredible edge over the house. Even more of an edge than the casino itself had against regular players.

After the hand was finished, Semyon leaned against the edge of the felt table, staring at the stacks of chips in front of him. Overall, in thirty minutes, they were up forty-four thousand five hundred dollars. The amount was staggering. He knew it was mostly variation from the mean; his expected win in that thirty minutes, given the two full shoes he'd played at a twenty-dollar per hand loss, and the one big round, was more like four thousand dollars. But if he kept at this, just using the single technique of cutting to the ace, he could get very, very rich.

He looked up from the chips, at the two pit bosses. The one with the white hair had gone back to his phone, and the bald one was strolling around the tables, an unconcerned look in his eyes. Even though Semyon and Allie had just jumped their bets to sixty grand in one hand, the pit bosses weren't suspicious at all. They had bought Semyon's act completely. And even if they hadn't—what had they seen? Two crazy high rollers throwing huge money down, seemingly at random. Even if the pit bosses had been trained to look for card counters, they wouldn't have seen anything that resembled card counting in Semyon and Allie's actions. They would never have guessed that Semyon had actually manipulated the cards to deal his own teammate an ace. It sounded too far-fetched, too impossible. And yet it hadn't even been particularly difficult. Nowhere near as hard as fixing a computer's hard drive or taking a high-level math exam.

Christ, this really was too fucking easy. Semyon looked at Allie, and she matched his eyes. He could tell that she was thinking the same thing. He smiled at her as they both lowered their bets back to their original thousand dollars.

"Well, *dyevchik,* are we ready for another hand?"

Allie flashed her eyelashes at him.

"Of course, Nikolai. The night is just getting started."

Victor was right.

They were going to bring the casinos to their knees.

CHAPTER 8

The Mirage
Las Vegas, Nevada

The bathrobe was some sort of silk/terry-cloth mix, with just a little too much sheen to it, a reminder that no matter how upscale you wanted to go, this was still Vegas. Likewise, the penthouse suite itself, for all its splendor and opulence, screamed of late-night parties with women who swung from brass poles for a living. The place was huge, at least fifteen hundred square feet, with two bedrooms, a dining room, two living areas, and three bathrooms. The carpets were thick, the walls done up in redwood and marble, and there were four separate chandeliers. The furniture was overdone, a mishmash of expensive antiques and contemporary pieces; most striking, however, was the preponderance of mirrors. There were more mirrors than places to sit, all of varying sizes and designs. Oval, rectangular, square, even triangular, hanging from walls and attached to ceilings, three in one of the bedrooms alone.

No matter where Semyon stood in the massive main living room, he could see himself reflected in one of the mirrors, and he couldn't help but feel self-conscious. The white bathrobe didn't help; nor did the long champagne flute he held in his left hand. He had never been in a suite like this before. He had never worn a bathrobe or drunk champagne from a proper champagne glass.

"I could get used to this," Allie said as she came out of one of the bathrooms, her own bathrobe swishing around her bare legs. "The champagne was a nice touch. We'll have to remember to send Kent a thank-you card when we get back to Boston."

Semyon laughed. Of course, they'd have to use unmarked stationery and have it mailed from a city like New York. Kent Tucker, the Mirage host who had set them up in the penthouse suite—on the recommendation of the two pit bosses—would find it very unlikely that two Russian high rollers would be visiting Massachusetts after the night they'd had at the Mirage.

One hundred and twenty thousand dollars. Semyon still couldn't get his mind around it. In just over five hours of play, he and Allie had nearly doubled the money they had dropped on the blackjack table. Playing one felt, just the two of them, using only the first technique. In that time, they had cut to aces seven times. Twice, they had been forced to use theatrics to get the dealer to show them the bottom card. Most of the time, the bottom card hadn't been useful, but luckily those shoes had gone favorably as well. Semyon knew that their expected win had actually been much lower—maybe twenty-five thousand. But they had gotten lucky. Extremely lucky.

Kent Tucker, the casino host, had nearly carried them bodily to the penthouse suite after he had gotten a taste of Semyon's bankroll. It was a host's job to make high rollers such as Semyon happy; if Semyon was happy, he'd keep coming back to the Mirage. Hosts were essentially independent contractors who built up stables of "whales"— players who were willing and able to gamble big. Tucker didn't care where the money came from, or even whether Semyon won or lost. Just that he was willing to bet big.

"Maybe we should throw in a Kalashnikov," Semyon joked back as he crossed to one of the many picture windows that lined the curved walls of the megasuite. "Or a gold-plated twenty-two. I think he'd appreciate that."

Allie laughed. Somehow, Tucker had gotten the idea that Nikolai Nogov was a Russian arms dealer. Semyon had never come right out and said anything of the sort, but when Tucker had questioned him about his business, he had made some offhand comment about recent travel to certain war-torn areas around the globe and a rather large shipment of "equipment" that had gone to an African nation in the middle of a civil war. Allie had nearly overturned her drink when Semyon had made the comment, but Tucker hadn't noticed. He had been too busy counting the avalanche of colored chips that Semyon had sent to the casino cage, to be exchanged for crisp hundred-dollar bills.

Semyon reached the window and looked out through the spider-web of reflected neon. He placed a palm flat against the cold glade of glass. Yes, Allie was right; maybe he could get used to this.

His thoughts were interrupted by a sudden buzzing, so loud that he jumped back from the window. He turned, looking at Allie. She shrugged, pulling her bathrobe tight against her throat. If they were about to get busted, they'd go down in terry cloth and silk. Semyon crossed toward the door. He was pretty sure that they hadn't done anything illegal, but he also knew that in Vegas, they could throw you out anytime they wanted. He doubted they'd do anything worse, but then again, if his past told him anything, it was that there was always the potential for getting your ass kicked.

He reached the door and glanced through the spyhole. To his surprise, the spyhole was covered from the other side. Someone had placed their hand over the lens. Semyon took a deep breath, prepared for the worst, and yanked the door open.

"I hope we're not interrupting the party. More important, I hope you've got another bottle of champagne and five more glasses, 'cause things are about to get a little crazy."

Semyon stepped to one side as Victor entered the suite, seemingly unfazed by the opulent decor. As usual, Victor was wearing one of his suits—navy blue, with oversize lapels and stiff white cuffs. He was also wearing wraparound sunglasses, even though it was now nearly 3:15 in the morning. He had a gray duffel bag slung over one shoulder.

"Victor," Semyon said, shaking his hand. "I thought we were going to meet at the volcano at four. How did you get past the security at the elevator?"

Kent had explained that you needed a special key card to access the penthouse suite. High rollers took their security seriously. But of course, Victor knew all there was to know about high-roller living.

Victor shrugged, pulling a handful of key cards out of his suit pocket and fanning them between his fingers. "Every VIP elevator in the city. I've stayed in most of them. I like this one a lot. And those bathrobes are divine, aren't they?"

"A pity they don't come in extra large." Jake's huge form filled up the doorway a few steps behind Victor. He smiled at Semyon, then stepped past him into the suite. His reddish beard was still bushy and unkempt, but he had traded his sweatsuit for an oversize Hawaiian shirt and tan khakis. "I mean, they call us whales, don't they? Fuckers could try and cater to the big-boned once in a while."

Female laughter trickled in from the hallway, and Semyon caught a strong blast of perfume as two blondes swept into the room behind Jake. He'd never seen either of the girls before—not during his training sessions, not before, during his short tenure at MIT. They looked almost like twins: short, busty, with matching, wavy platinum locks. The first girl was wearing a miniskirt, a black velvet halter top, and superhigh heels. The second was dressed barely more conservatively, in jeans and a leather bustier.

"Don't brag, honey," quipped the one in leather. "I wouldn't exactly call you big-boned. Average-boned, more like."

"Now that's just vicious," exclaimed the one in velvet.

The two girls smiled at Semyon, and Victor made the introductions.

"Misty and Carla, this is Semyon Dukach. Semyon, I know they look like sisters, but they're actually just roommates, separated at birth. Seniors and engineering majors, though they don't look it. They've been playing with me for about a year. They make an interesting addition to our team, don't you think?"

Semyon had to agree. The two blondes were about as unlikely a pair of professional cardplayers as one could find. He watched as Carla—the one in the leather bustier—followed Jake onto one of the couches, across from Allie. Obviously, the pretty senior and the huge Harvard law student were more than teammates. The other blonde, Misty, moved next to them on the couch as Allie began searching for more champagne glasses. Semyon turned back around in time to see one more player enter the suite. This time there was no need for an introduction.

"Shit, man, a few hours in Vegas and you've already gone Hefner on us."

Semyon grinned as Owen tossed a bright red duffel past him and sauntered into the suite. Owen looked worn-out from the five-hour flight, his hair sticking up from his head like a demented halo. He gave Victor a wide berth as he crossed the room to the window, checking out the view. Semyon knew that Victor had been extremely reluctant to add the rebellious math major to the team, but Semyon was glad Owen was with them. Besides, he was damn good with the cards, nearly as good as Semyon. He'd be an asset, of that Semyon was sure.

By the time Semyon had shut the door and joined the rest of the team at the couches, Allie had found enough champagne glasses for everyone. Victor looked like he was about to make a toast, then thought better of it, simply downing his champagne in one swift motion. Semyon looked around the room, at what seemed to him a very strange, almost impossible cabal. A slick young mastermind in an expensive blue suit. An overweight law student draped in an obscenely garish Hawaiian shirt. Two pretty young coeds, who looked more like strippers than engineers. An enigmatic rocker in a leather jacket. Allie, gorgeous, ambitious, and smart. And Semyon himself, a computer geek pretending to be a Russian arms dealer.

There really wasn't any need for a toast because they were all probably thinking the same thing.

CHAPTER 9

The Flamingo
Las Vegas, Nevada

his is completely insane."

Victor just smiled as Semyon pulled Owen's leather
jacket over his bare shoulders. He went to zip up the front,
but Victor stopped him halfway up his naked stomach.

"Leave it partly open. We gotta make this realistic."

"Realistic?" Semyon blurted. He glanced at himself in the bathroom mirror. Hair slicked back, tight jeans pulled down low on his
waist, some of Allie's dark eye shadow applied liberally under his
eyes—and Owen's jacket, open all the way down to his last rib. There
were goose bumps on his pale chest from the air-conditioning, and
he felt extremely self-conscious without his shirt. But Victor smiled
again, slapping a hand on his shoulder.

"I'm telling you, it works. Rock-and-roll, man."

"Yeah, sure. How come you get to wear a suit?"

"Because I *am* the suit." Victor grinned. He straightened his
thin silver tie in the mirror. Semyon had to admit, Victor at least
looked the part. He had placed a pair of wire-rimmed glasses—
nonprescription lenses, since his vision was perfect—over his eyes
and had added some gray flecks—baby powder—to his dark hair.
With the perfectly tailored suit, he looked a good ten years older
than when he had first walked through the door of the Mirage suite.

"Come on, don't tell me you never wanted to be a rock star. Now you get to live the dream."

Victor turned away from the mirror and started toward the door. Semyon took one last look at himself, then shrugged. They had believed him as a Russian arms dealer. Why not a European pop star, making his first trip to the United States?

Victor held the door open for him, and they exited the bathroom. They found themselves at the edge of a wide bank of slot machines, blinking, beeping mechanical monsters that seemed to stretch on forever. The air was smoky and dim, the ceilings a little too low. They had entered the Flamingo from the Strip, but had wound their way beyond the crowded front entrance to the quiet slot alcove to finish Semyon's transformation. It seemed fitting, to him, that Victor had chosen the Dirty Bird as the first casino they were going to hit as a team. In essence, the Flamingo had been the first real megacasino built in Vegas, way back in the 1950s. At the time, it was known for the pink flamingos that had been placed around its lavish pool and the cavalcade of celebrities that Bugsy Siegel—the infamous mob boss—had lured to his oasis in the desert. A "carpet joint," which was supposed to take the glamour of Hollywood and transport it to the burgeoning capital of sin. *What better place for the improv scene Victor and his crew were about to attempt?*

Victor waved Semyon ahead. Semyon had barely taken one step forward when he suddenly felt hands all over him—followed by thick blasts of fruit-flavored perfume.

"Oh my God, you are so hot," squealed Carla as she kissed his right cheek. Misty had her arms around his waist, squeezing him from behind. "Let's go win some money, baby."

The two girls had reshellacked their already heavy makeup with bright colors that made them look even younger than they really were. Misty had pulled one shoulder off of her velvet halter top, revealing more of her ample chest. Carla had piled her blond hair high on her head and had added a pair of garish chandelier earrings. Both were chewing gum.

Semyon laughed inwardly as the two girls walked him through the banks of slot machines. Victor was a step behind and had been joined by Owen, who was now also wearing a blue suit—though of lesser quality than Victor's—and had taken some control of his spiked hair. Allie moved next to them as they passed the next bank of slots, her own halter top hidden beneath a conservative white blazer. Jake had remained behind in the hotel room, so for now, it was just going to be the six of them. *A Euro-pop star, two groupies, a pair of managers, and their assistant.* When Victor had pitched the characters to Semyon, he had thought it unlikely they could pull it off; but glancing back over his shoulder at the crew spread out in his wake, he had to admit it looked pretty authentic. Anywhere else in the world, it might have seemed improbable, but here in Vegas—well, it actually *worked*.

They turned a corner and found themselves in the main section of the first-floor casino. Table games spread out ahead of them, and even though it was now a few minutes past 4:00 A.M., the place was hopping. As a casino, the Flamingo was not anywhere near as posh as the other megaresorts on the strip—the Mirage, Caesars, the Sands—but its mid-Strip location, its pool, and its history made it a resiliently popular choice for the weekend crowd. And no matter how downscale the Dirty Bird had become, in terms of gambling, there was always a place for high-roller action at the former gangster's paradise.

Victor touched Semyon's shoulder, pointing past the crowded tables toward a velvet-roped-off section just past the entrance to the buffet. Semyon nodded, leading them toward the Flamingo's high-stakes area. As they moved forward, Semyon noticed people turning and looking. Some even pointed, making comments about the two girls on Semyon's arms, wondering aloud who the fuck Semyon could be. Semyon ignored the stares, his mind focused on the velvet rope.

They reached the rope and Semyon paused, letting Allie move quickly in front. She held the rope to one side so the group could pass through. She winked at Semyon as he moved past, and he felt the heat rising into his shoulders. He was nervous, but he was also excited. *This was going to be fun.*

There were ten blackjack tables in the high-stakes lounge, as well as two craps tables and a roulette wheel. The place was fairly empty—maybe a dozen gamblers at most—and the decor was nowhere near as lavish as at the Mirage. Paintings of flamingos on the walls interspersed with portraits from old Hollywood. Thick green carpeting, a few palm trees in the corners. There wasn't a dedicated bar in the room, but at least four cocktail waitresses worked the tables. There were three pit bosses in the center of the room, and all three sets of eyes were already on Semyon and his team as they made their way past the first blackjack felt. Semyon fought back the sudden moment of terror that erased the warmth from Allie's wink; were they really going to buy this act?

Before Semyon could answer his own internal question, Victor was tapping his shoulder, pointing to a table by the edge of the room, right beneath a framed photo of James Dean. Instinctively, Semyon's eyes moved to the dealer.

Male, midfifties or early sixties, with thick gray hair and great big bags under his eyes. His lips were downturned, his shoulders thick and slightly hunched beneath his light green Flamingo uniform. His expression was tired, bored—old Vegas, he had seen it all before. Hell, with those hangdog eyes and wrinkled skin, he looked like he'd been at the Flamingo since the day it opened.

Semyon shifted his attention to the man's hands. Gnarled, oversize, like paws with fingers. They flung the cards toward the single player—a young Asian man with thick glasses and a bowl haircut—with military precision, and it was obvious the dealer had been working the blackjack felts for a long time. *An old-timer, a lifer.* With big, perfectly measured hands.

Semyon smiled inwardly. The old-timer was almost the exact opposite of the dealer he and Allie had worked earlier in the evening. Victor wasn't planning on using his first technique at the Flamingo. He had something else in mind.

Semyon turned and led his crew toward the portrait of James Dean. He tried to ignore the three pit bosses, who were also now converging on the table from the other side. He knew that he was going to have to get used to the sudden attention because that was what Victor's game was all about. If all eyes weren't on them, they weren't doing it right. This wasn't a clandestine operation, this was right out in the open. Like Vegas itself, the bigger and brighter, the better.

He reached the table and chose the empty seats farthest away from the Asian gambler—third base. He hit the seat hard, throwing his legs over both sides in a cowboy straddle. Carla and Misty stayed with him, hands on his bare chest, jabbering while violently smacking their gum. Immediately Victor and Owen took the seats to his right, Allie next to them. Allie already had two handfuls of banded hundred-dollar bills out on the felt and followed them up with four more. Then four after that. By the time she reached two hundred thousand, the dealer had stopped dealing, and the three pit bosses had reached the table. At the same time, the Asian guy with the thick glasses took one look at Semyon and the two girls hanging from his shoulders, scooped up his feeble collection of hundred-dollar chips—and stumbled away, leaving them the table. *Perfect.* Victor hadn't even needed to have them raise the table limit.

Allie pushed the bills forward as Semyon watched the old dealer take the cards and stack them in preparation for his shuffle. The guy's name tag identified him as Bruce, no last name. The minute he started splitting the stacked cards up for his shuffle, Semyon knew Victor had made the perfect choice for his second technique. Bruce barely looked at the cards as he worked, stacking the cards in perfect blocks, then riffing them together with the precision of a machine.

Semyon raised his attention back to the three pit bosses, who were trying to decide who to talk to first—Allie, with the money stacked in front of her, or Victor and Owen, who by their suits seemed most likely to be in charge. For the moment, they seemed content to wait for the dealer to finish so he could exchange the bills for chips. Two of the pit bosses looked pretty fresh, maybe in their early thirties, but the third looked almost as old and hardened as the dealer. Wide-shouldered, with a jutting square jaw and a smile that looked like it had been applied—painfully—with a scalpel.

When the dealer finished with the cards, stacking them back into the shoe, he reached for the money in front of Allie. As he exchanged the cash for chips, Victor pointed a finger toward the hardened pit boss.

"We're gonna need champagne, and lots of it. Our boy here is celebrating his U.S. debut. Dietrich, give 'em a taste of those pipes."

It took Semyon a second to realize that Victor was addressing him—and another second to realize what Victor wanted him to do. Semyon's eyes widened as Misty clapped her hands.

"Come on, Dietrich baby. Sing the one you wrote for me. You know, the one that goes real high-pitched at the end."

Semyon could feel all three pit bosses watching him, as well as Victor and his team. *That fucker.* Well, Semyon knew he didn't have a choice, he was going to have to play the part. He took a deep breath and started singing—in a truly awful German accent.

"Baby you are the one, you are, you know, the one . . ."

Thankfully, Victor cut him off before he had to invent a chorus.

"So you can see why we're ready to freakin' celebrate!"

The dealer had finished with the chips, and Victor deftly split them among the players. Meanwhile, Allie handed the pit boss a stack of ID cards. Semyon hadn't seen these yet, but he assumed that "Dietrich" would soon be getting the kind of comps at the Flamingo that Nikolai Nogov was enjoying at the Mirage. The pit boss with the jagged mouth barely looked at the pictures on the cards; he was too busy watching five-hundred-dollar chips being pushed into all six of the betting circles at the table. Victor and Allie were each playing one circle, while Owen and Semyon had taken two each, rounding out the felt. They were starting with the table minimum because as with the first technique, the second technique was opportunistic. The opportunity for a real advantage wouldn't come for a while—but when it did, as with the first technique, it would be fantastic.

Bruce, the dealer, began his deal, his thick hands sliding the cards mechanically across the table. Allie's hand came first, a king of diamonds, a good starting card, but that was only of tangential concern. All of them would play perfectly, but that wasn't how they were going to make their killing. Perfect play still left the casino ahead by almost 2 percent. To win, they had to be more than perfect.

Victor got a four of hearts. Owen, a jack of clubs on his first hand, a three of hearts his second. Semyon's first hand started unremarkably, a seven of diamonds, but on the last circle on the felt, he got dealt an ace of clubs. He smiled inwardly, glancing at Victor. Victor was still jawing with the pit boss; he had launched into some story about their flight up from L.A. in a private jet owned by one of the music studios. He didn't even seem to be paying any attention to the cards, which were still coming—but Semyon knew better. Victor was just *that* good. He had certainly seen Semyon's ace of clubs.

The dealer continued to deal. He gave himself his down card, then started again with Allie. A queen, giving her a powerful twenty—not that it really mattered. Then a nine for Victor. Owen received a six and a ten, and then it was Semyon's turn. On his first hand, he got a queen of hearts. And on top of his ace, a nine of diamonds. So reading his cards backward: the ace of clubs, a nine of diamonds, a seven of diamonds, and on top of that, a queen of hearts.

Semyon blinked hard, and in that instant the four cards became words—words that he'd memorized previously from a chart Victor had made him carry around with him. *Cat*: the ace of clubs. *Dip*: the nine of diamonds. *Dog*: the seven of diamonds. And *Queen*: for the queen of hearts. It took him another instant to reverse the four words: queen, dog, dip, cat—which he then turned into a quick little story, one strange enough that it would be easy to remember. *The queen and her dog took a dip with the cat.* By the time his blink was over, he'd used the mnemonic to put the four cards to memory. He knew that Victor, Owen, Allie, and even Carla and Misty had all just done the same.

Now all of them had the four cards marked in their heads. Even as they played out the hand—hitting when they were supposed to, splitting and double-downing, cursing when they lost and shouting when they won—that little story remained in the back of their minds: *The queen and her dog took a dip with the cat.* Queen of hearts. Seven of diamonds. Nine of diamonds. Ace of clubs.

The dealer finished dealing the round and quickly scooped up the cards. Right to left, the same rote scoop that every dealer in Vegas— every dealer in the world—used to clean up a blackjack felt. Semyon watched as his ace of clubs was taken first—followed by the card on top of it, the nine of diamonds. Followed by the seven of diamonds. And then the queen of hearts. Now they were stacked, from top to bottom, exactly in the same order he had memorized them, the exact order of his little mnemonic story. The dealer continued down the felt, scooping all the rest of the cards. Then he placed the stack into the discard pile.

He began dealing the next round, but by then Semyon was barely paying attention. Because what really mattered were not the cards being dealt; what mattered were the cards that had already *been* dealt. Sitting there, in a pile in the discard rack, in the exact order that he—and the rest of his team—had memorized.

His excitement rose as he continued to play; hand after hand, deal after deal, as Bruce dealt deeper into the shoe. The discard stack kept getting higher and higher, but still, at the bottom, were those cards from the first round. *The queen and her dog took a dip with the cat.* Queen of hearts. Seven of diamonds. Nine of diamonds. Ace of clubs.

When Bruce had finally finished out the shoe, stacking all the cards in the discard rack to get ready for the shuffle, Semyon was down about a thousand dollars. Victor and Owen were both down as well, but Allie was up a bit—so together, the team was down a few thousand in total. But Semyon's excitement had grown exponentially—he had a good feeling that things were about to get much better.

He watched as the dealer rolled the cards for the shuffle, splitting the stack into two even piles. Semyon could still see the cards from the very first round, stacked together at the bottom of the pile at the dealer's right hand. As the dealer worked his way down the two piles, Semyon kept his eyes on that part of the deck.

Finally, the dealer reached the bottom two piles and riffed them together with his oversize, overtrained hands. The riff was perfect—the cards floating together in an exact shuffle. The thing was, even though the cards had been shuffled together, the sequence that Semyon had memorized still existed—just separated by the cards from the second stack that had been shuffled into the first. And since the old dealer's hands moved so mechanically, making his shuffle so perfect—Semyon knew that those cards were each separated by a single riffed card. A garbage card, a random card that didn't matter—because the sequence still existed, exactly as Semyon had memorized it. *The queen and her dog took a dip with the cat.* Queen of hearts, seven of diamonds, nine of diamonds, and an ace of clubs.

Bruce, the dealer, moved the cards into another pair of parallel stacks and began his shuffle again. And again, Semyon's sequenced cards got riffed with a garbage stack—and again, the riff was perfect, the shuffle exact. Each sequence card, and each random card, got one more random card added in between. The sequence still existed—but now it was broken up by three garbage cards in between each memorized card. *The queen and her dog took a dip with the cat*—with three random cards in between each step.

The dealer rerolled the cards, then held out the cut card toward the table. Victor grabbed it before anyone else had the chance—not that any of them would have challenged him for the honors. He chose a spot about two decks in front of the sequenced cards—and smiled as Bruce restacked the shoe.

"I think this is gonna be our lucky round, eh, Dietrich?"

Semyon smiled back at him. "*Achtung,* baby."

The two younger pit bosses had already wandered off, convinced that the bizarre crew were simply fools on their way to being parted with some major pop-rock cash. But the hardened pit boss was there to offer a patronizing smile at Victor's exuberance. He watched as the six five-hundred-dollar bets came back, and the deal began.

The cards ran quickly through the first two rounds, and Semyon fought hard to control his nerves as he kept track of the cards. He wasn't sure how he could remain so cool on the outside, with his insides churning, but apparently outward patience was a virtue he'd inherited from parents who'd suffered through years under a Communist dictatorship. He and the rest of his team played out their hands by rote—with a slight variation on basic strategy. They were extremely careful about how many cards they took; in the first round, Victor stood on a twelve against the dealer's seven, and Owen stood on a fourteen when he, too, should have hit. The truth was, the hands themselves were not as important as the number of cards they took out of the shoe; they were counting *down*—not counting *cards*—and that made all the difference.

The cocktail waitress arrived just as the dealer was sweeping away

Semyon's second hand—and a lost five-hundred-dollar chip. The waitress handed him a crystal champagne flute. No plastic cups in the high-stake's lounge, even at the Flamingo. Semyon turned to toast Victor, who winked at him.

"I think we should get a little crazy now," Victor said, grinning, and Semyon immediately understood. He'd been counting down the cards as they came out and knew they were closing in on the sequence they'd memorized.

Semyon grinned back, grabbing stacks of chips with both hands and suddenly pushing them into the betting circles in front of him. *Two hands of ten thousand dollars.* Owen raised his eyebrows, then suddenly did the same. Victor and Allie remained at their original five-hundred-dollar bets. Their hands were going to be buffer hands—in case the sequence somehow came out a little early.

The pit boss shrugged at the dealer, who began dealing out the hand. Semyon watched as the first card came sliding out of the deck. *Bam.* The queen of hearts. *Queen.*

Of course, there was always a chance it was a coincidence. There were six queens of hearts in the six-deck shoe. But given how Victor had cut the deck, and how Semyon had been expecting the sequence to occur—he was pretty certain it was *Queen*. Now they just needed to tell the rest of the story.

Victor was next. He was dealt a four of spades. Then Owen's hands. A six of clubs and a ten of hearts. Then came Semyon's first hand. He watched the dealer's mechanical arc as the card spun toward his betting circle—the seven of diamonds. *Dog.*

Halfway there, Semyon thought to himself.

Semyon's second hand was a ten of clubs. Then the dealer got his down card. The dealer moved back to Allie, who drew a nine of spades. Then he shifted to Victor—and there it was, the nine of diamonds. *Dip.*

The dealer was now at Owen's first hand. On top of his six of clubs, he drew a ten of hearts. Then his second hand, a king of spades. Semyon's first hand received a four of hearts.

He braced himself, the mnemonic running through his head as the final card came out.

Cat.

The ace of clubs, landing right on his king of spades, for a natural twenty-one. *Blackjack!*

Semyon clapped his hands in real joy. Victor slapped the table with both hands, then shouted at the dealer: "Now that's a pretty sight! The Snapper!" Both Carla and Misty were jumping up and down—a sight that was drawing gazes from the nearby tables. But the dealer simply shrugged and paid out Semyon's blackjack win—with three five-thousand-dollar chips.

Fifteen thousand dollars, on a single hand. And it hadn't been luck. It hadn't been chance. Semyon, Victor, and the rest had tracked a sequence of cards through the shuffle and had used that sequence to manipulate an ace into Semyon's ten-thousand-dollar bet.

Key-card sequencing, Victor had named it—the second technique. And it was truly brilliant. They had carried it out right under an experienced pit boss's gaze, right in front of an experienced dealer. Unlike card counting, the technique was impossible for the casinos to see. No matter how many cameras were trained on them, no matter how much heat they drew. And of course—the technique was perfectly legal. Victor hadn't tampered with the game itself, he hadn't physically marked any cards or used any sort of computer technology. Just some math, some observation, some acting, and a whole lot of practice. As with the first technique, Semyon had a 51 percent edge on the hand with the ace. The rest of the hands were at a small disadvantage, but overall, it was a powerful edge over the house. It was something they could continue doing all night long; memorizing sequences that led to aces, then playing out the hands to get those aces onto the big-money bets.

"You see, Dietrich?" Victor said, leaning over Owen to pat Semyon on the shoulder. "I told you that America was a land of opportunity. And that goes triple for Las Vegas. Isn't that right, Bruce?"

The hardened dealer grunted at Victor, then began to deal the next round. Semyon was sure the old-timer thought he'd seen it all before; young hotshots who thought they could beat the system, thought they could get lucky and walk away winners.

Semyon wondered what the old-timer would say if he somehow

discovered that at that very moment, all six of the players in front of him were memorizing another sequence of cards that led to a powerful ace, mentally repeating seemingly nonsensical words that would soon lead to another massive payout.

Hell, he'd be forced to change his entire worldview. Because the truth was, the system *could* be beat. In just the past few hours, Semyon had already used two of Victor's powerful techniques on two unsuspecting casinos—and had won big. Semyon had already beaten the system twice—and his night wasn't over yet.

Victor's third technique—the most powerful technique of all—was still to come.

CHAPTER 10

A Casino on the Strip
Las Vegas, Nevada

Six A.M.

Strolling down a concrete walkway, still hopped up on the energy of a city that had no real sense of time. Eyes blinking hard as they scanned the sky above the enormous stone-and-marble structures, the rising sun bright against the careful, chiseled curves of the massive statues. The neon from the Strip was brighter still, reflected off the glass entrance to the huge, Disneyfied casino that sprawled out ahead.

"Now, *this* is a casino," Semyon exclaimed as he and Owen quickly took the last few steps up the walkway toward the megaresort's front entrance.

"We who are about to kick ass salute you," Owen responded, smirking as he let Semyon step through the glass doors in front of him.

Semyon grinned as the brightness from outside was replaced by the familiar dark air of a bustling casino. The ornate design features of the exterior seemed much less prominent inside, amid the mind-numbing barrage of row after row of blinking, beeping slot machines. Off to the right, arched hallways led visitors toward the upscale shops, but Semyon and Owen hadn't come here to shop. They had chosen this casino because it was one of the best places for blackjack on the Strip and could handle high-roller action; meaning, like the Mirage and the Flamingo, this casino didn't balk at ten-thousand-dollar bets—or million-dollar players.

Semyon and Owen wound their way past the slot machines toward the table games that lined the middle of the casino. Even though it was six in the morning, the place was still pretty crowded. Every other city in the world was just waking up, but the truth was, Vegas hadn't yet gone to sleep.

"You'd never guess the sun's coming up outside," Semyon commented. "No windows, no clocks. I bet more people miss their flights here than anywhere else in the world."

"Fuck yeah," Owen responded. "Time's just an inconvenience to the casinos. If the people who built this place could figure out how to get rid of the sun, they'd have done so already. Replace it with one big neon tube, let it run forever."

Semyon laughed, but part of him believed Owen wasn't so far off. The people who ran these casinos didn't want you thinking that it was 6:00 A.M., that one day was ending, another starting. They wanted it all to blend into one massive binge, one marathon gambling session that ended only when your wallet was empty.

Semyon slowed his pace as they reached the edge of the blackjack pit. To his surprise, he saw a handful of hundred-dollar tables right at the edge of the gambling area, and even a few five-hundred-dollar tables. He guessed that the casino was making a push for the high-end players on this crowded Vegas weekend. If he and Owen found what they were looking for, they'd be happy to oblige.

He began pacing around the pit, Owen right behind him. What came so naturally to Victor—spotting the necessary components for the techniques—took a little more effort for him, but he was sure that would change over time. Even now, just hours since he'd landed in Vegas, he felt more comfortable in a real casino setting. He wasn't anywhere near as nervous as he'd been when he and Allie had first hit the Mirage. Although, of course, he wouldn't have minded having her by his side again. Victor had decided it would be better off for them to split into three groups to cover more ground as they moved on to the third technique—which needed a team of only two to pull off. Allie was with Victor, hitting the Stardust, while Carla and Misty had rejoined Jake at the Mirage. Semyon was actually glad that Victor had chosen to play with Allie instead of teaming with him; he wasn't sure why, but he was still uncomfortable around the blackjack guru. There was something about Victor he still couldn't read; maybe it was just that even after three months, he knew so little about the dapper young genius; he had nothing to build an impression on. And he was also glad not to be teamed with Jake and the girls because he'd have felt like a fourth wheel; there was no question that something heavy was going on between the Harvard Law behemoth and Carla, and Misty was really just an extension of her roommate, an extra component to her bouncy personality.

That left Semyon with Owen, which was just fine, since the two of them worked well together. They had even practiced their characters a few times at Foxwoods, the Indian-reservation casino in Connecticut: Dr. Michael Nussbaum and Dr. Irvine Powell, young dentists in town for a convention. Semyon had traded Owen's leather jacket for a button-down oxford shirt, fastened all the way to the neck, and Owen had remained in his suit, changing only the tie to something a little more conservative. Hair parted severely, walking with the stiff gaits of young men who'd gone to Northeastern prep schools, they looked young, rich, and boring. Nobody was going to bother a couple of DMDs blowing root-canal profits at a blackjack table.

"Yo, Doc Nussbaum." Owen coughed as they worked their way through a group of college kids in football jerseys, cheering around a craps table. "I think we might have some luck over there."

He was pointing toward one of the five-hundred-dollar blackjack tables just beyond a crowded roulette wheel. The table was empty, the cards still unshuffled and spread out on the felt. The dealer was a man, young—maybe even as young as they were—and most important, short. Five four at most, with small features, a sharp, triangular chin, and narrow, ferret eyes. He looked bored, standing there by himself at six in the morning, and Semyon could imagine the thoughts floating through his diminutive head. He wanted out of the graveyard shift, the sooner the better. As Semyon shifted his gaze to the little man's hands, those miniature palms lying flat against the felt, he was glad that the man hadn't yet been granted his wish. *Those spindly fingers were perfect.*

Semyon nodded at Owen and moved quickly toward the empty table. He chose the first-base seat, right up near the dealer, and before sitting, he made a big show of leaning over the table to read the dealer's name tag.

"Hey, Bobby, my friend and I need to kill an hour before our flight back to Cleveland. You're going to be kind to us, aren't you?"

Bobby, the dealer, smiled, showing one of the worst overbites Semyon had seen in a while. Owen didn't miss a beat as he dropped onto the stool two seats over from Semyon.

"Bobby, man, you should get those chompers looked at. Here's my card, got our office number right at the bottom. Give us a call if you're ever in town."

He tossed the dealer an ivory colored business card, along with a pair of work IDs with his and Semyon's pictures on them. Before the dealer could pick them up, Semyon had a hundred thousand dollars out on the table, in neat stacks of banded hundred-dollar bills. The dealer froze, then turned and searched the pit behind him for a pit boss.

It took a good few minutes before an oversize, middle-aged woman with frizzy blond hair strolled over. Her skin had an unhealthy gray pallor to it, the signature of a lifelong smoker. She had serious gray eyes, a splash of orange red color on her wide cheeks, and a thin red line above her chin that bore a mild anatomical similarity to a smile. Her name tag identified her as Dorothy, and she looked like a Dorothy—one who'd been away from Kansas a long, long time. She was the first female pit boss Semyon had seen since he'd arrived in Vegas, though something about her seemed even more old-school than the hardened dealer from the Flamingo.

"You boys having a good time?" she said, taking their IDs from the dealer. "DMDs. That's dental medicine, right?"

"That's right, sweetheart. Biggest tooth-and-gum outfit in Ohio. I've filled enough cavities in five years to pave over the Grand Canyon."

Owen was really on a roll. If Semyon hadn't known better, he would have believed that his friend really was drunk—or worse. The whites of Owen's eyes had a reddish tint to them, and his hair had gone a little wild again, errant locks sprouting out like sunbursts. He had visited the bathroom on their way out of the Flamingo, so he'd obviously done something to make himself look more the part. Victor had taught them that a couple of drops of vodka in your eyes and a splash of Scotch on the front of your shirt told almost as good a story as a fake ID, but Semyon wasn't really sure what Owen had done to make himself over. In any event, the act was working on Dorothy just fine.

"Glad to have you, Doctors," she said, nodding at Bobby to start shuffling. "When you're done at the table, I'll make sure to have one of our hosts come by to tell you about our beautiful suites."

She stepped away from the table, but Semyon was pretty sure she'd be keeping her eyes on the two highest rollers in the pit. Bobby quickly began shuffling the cards, his spindly little fingers moving quickly through the ritualistic routine. He wasn't as mechanical as the old-timer from the Flamingo, but he was competent, and fast. So fast that he'd had the cards rolled over horizontal onto the felt before Semyon had even finished stacking half the one hundred thousand in chips in front of himself, half in front of Owen. The thing was, although the dealer was fast, he was also sloppy; his fingers were too thin to properly cover the back card on the rolled deck, and as he held out the cut card toward Owen, he flashed the back of the deck almost directly at Semyon. To a dealer, it was a meaningless thing; what did it matter if a player saw that last card? The cameras upstairs could catch it again and again, and still no one would even think of admonishing the dealer for his sloppy role. But to Semyon and Owen, it was like cash from heaven. A flash of bright red card was clearly visible, along with the edge of a curved letter in one corner. *J*.

A jack, either hearts or diamonds.

Perfect.

Semyon had expected to spend a while at the casino before he'd have a chance to try the third technique, but there it was, right in front of him, the opportunity he was looking for.

He reached forward and grabbed the plastic yellow cut card before Owen could have a stab at it. He wanted to do this himself; an initiation of sorts, because the third technique wasn't just the most powerful of all three techniques, it was also in some ways the trickiest. Although the key-card sequencing that they had pulled off at the Flamingo was fairly advanced, there was actually room for error. Equally, with the ace cut he had done with Allie, it didn't really hurt to miss by a few cards; as long as the ace landed on one of your six hands, you were going to come out ahead. With the third technique, it was about perfection.

He waved the yellow plastic cut card in the air and, with a flourish, jabbed it right into the rolled stack of cards. In his head, he saw the fifty-second and fifty-third card from the bottom cleaved apart, an exact single-deck cut. He didn't wonder for a second if he had missed his mark; he knew, in his soul, that he had cut exactly fifty-two.

The dealer moved the cards to the back of the deck—putting the jack that Semyon had seen exactly fifty-three cards from the top of the stack—then moved the entire lot into the shoe. Semyon looked at Owen, and in one motion, they each filled three betting circles with single five-hundred-dollar chips.

As with the first technique, the first round of play didn't really matter. Semyon watched as the dealer quickly dealt the cards, counting down along with each twist of the man's spindly hands. By the end of the round, seventeen cards had come out of the deck. In his head, Semyon quickly did the subtraction. Fifty-three minus seventeen was thirty-six. He and Owen replaced their lost bets, keeping five hundred-dollar chips in each betting circle. The next round began, and again the cards came out. Including the dealer's two, the deal brought out fourteen more cards—meaning now there were twenty-two cards left before the jack.

Semyon quickly glanced over his three hands and Owen's three hands. He had two twenties and a fourteen. Owen was showing an eleven, a fourteen, and a twelve. The dealer had a ten showing. According to basic strategy, Semyon knew he was supposed to hit his fourteen, while Owen would double-down on the eleven, hit the fourteen and the twelve. That would mean at least four hits, four more cards taken from the shoe—which would bring the total amount of cards left until the jack to eighteen.

Not bad, he thought to himself. They didn't even have to vary from basic strategy for the manipulation to work. They took their hits, the dealer turned over his hidden card to make a twenty, then swept the cards off the table. It was time for the third round.

Semyon smiled widely at Owen. He didn't need to say anything. Without pause, the two of them both pushed three stacks of ten thousand dollars in chips into their betting circles. *Sixty thousand dollars onto the table at once, the maximum allowed.* It was a massive amount of money. Semyon felt a shiver move through him as the dealer looked back at the female pit boss, who shrugged her thick shoulders.

The cards came out. Semyon pulled a seventeen, a twelve, and a twenty. Owen also got a twenty, a nine, and an eighteen. The dealer had a seven showing. That was fourteen more cards out of the deck, which meant that the fourth card left to be dealt was the jack Semyon had seen.

Now it was just a matter of him and Owen playing the hands. He knew that what they were about to do would look a little bizarre. But it didn't matter. Even if he tried to explain his play to the pit boss, he bet she wouldn't understand. She just wouldn't have the math to understand.

With a smile, Semyon pointed to his seventeen and asked for another card. The dealer raised his eyebrows. With ten thousand dollars down, Semyon was hitting on a seventeen against a seven. It was as stupid a move as one could make. Except, Semyon knew something the dealer didn't. He knew that in four cards, a jack was coming.

The dealer gave him a five, busting his hand. He winced instinctively as Bobby scooped up his ten thousand dollars. Then he pointed to his twelve, asking for a card. The dealer gave him an eight, for a nice twenty. Semyon held on his last hand. He'd taken two cards, which meant the jack was now two cards away.

Owen grinned, pretending to survey his own cards. Then he shrugged, pointing to his nine. "Gimme something I can use."

The dealer handed him a four, for a thirteen. Owen faked a grimace, but inside, Semyon knew he was as thrilled as Semyon was. Because if they had counted down wrong, or had made a poor cut, that jack could have come out early. But it hadn't. It was still in the deck. And if Semyon had done everything right, it was the next card to be dealt.

The dealer turned over his hole card, revealing an eight, for a fifteen. Semyon's heart slammed in his chest as the dealer went for that next card and pulled—the jack.

The jack.

"Dealer busts!" Bobby said reluctantly.

Semyon looked at Owen.

"Pay the table!" they shouted.

The third technique had worked perfectly. Semyon and Owen had steered a ten card—a bust card—to the dealer. They had purposefully busted his hand. Although they couldn't have known what his hole card was, mathematically, a ten card steered as the dealer's third card—or draw—was worth, on average, 30 percent *per hand*. That meant, 30 percent for each of Semyon and Owen's six ten-thousand-dollar hands, or eighteen thousand dollars *expected* return. And since there were so many ten cards in a six-deck shoe, the odds of having the opportunity to use the third technique were actually quite high.

It was a staggering advantage. A guided ace, as in the first technique, gave the player a single hand with a 51 percent advantage. The key-card sequencing of the second technique worked out to the same advantage, 51 percent for a single hand. And a ten guided to one hand would be worth only about 13 percent. But a ten card guided to the *dealer,* that was unbelievably powerful. *Thirty percent per hand.* More than any card counter in the world could earn. In fact, more than the casino itself could be expected to earn against an average player. At blackjack, the casino had a 2 percent edge over a perfect basic-strategy player. At roulette, the worst game in the house, the casino won about 5 percent per bet. Two percent, 5 percent. Even slot machines varied between a 5 percent and 15 percent edge. But the third technique, busting the dealer, gave Semyon and his team a massive *30 percent* advantage.

And the technique was so subtle, so unheard of, that no casino in the world would ever be able to figure out what they were doing. They certainly weren't cheating. Cheating, in Semyon's mind, was changing the outcome of the game. He was simply using the information in front of him to beat the game. Sure, he had seen the back card—but he hadn't done anything illegal to make it visible. And they had *handed* him the cut card. Was it wrong for him to cut to an exact place in the deck? Was it wrong for him to use the fact that he knew where that jack was ahead of time? Maybe the casinos would see it differently, but Victor had simply figured out a way to turn these facts—these observations—to their advantage. To their *massive* advantage.

The dealer began paying them out. Minus Semyon's one overdrawn hand, they had earned fifty thousand dollars by busting the dealer. Fifty thousand dollars in a single hand. In less than ten minutes of play.

"Unbelievable," Semyon murmured aloud.

With what they had won at the Flamingo and the Mirage, the team, as a whole, was already up well over two hundred thousand dollars on their first night in Vegas. More money than Semyon's father had made in his lifetime, more than Semyon could make repairing a thousand computers or working five years in an MIT lab. And the weekend was just getting started. God only knew how much they could win in forty-eight hours of play. No matter how strange or hard to read Victor was, the guy was a true genius. Semyon had been lucky as hell to stumble on that poster in the Infinite Corridor.

Make Money Playing Blackjack.

Fuck, that was the understatement of a lifetime.

Semyon grinned as he pulled back his winnings. He was about to start restacking the huge pile of chips in front of him—when suddenly he felt a hand on his shoulder.

He turned—and froze.

There were three burly security guards standing in a tight semicircle behind him. Three more were surrounding Owen, who was already half out of his seat. The female pit boss was coming around the blackjack table, followed by two more pit bosses, both male, one of whom was talking on a cordless phone. By his manner and demeanor, the one with the phone looked like he was in charge. Gray-haired, thin-lipped, with hulking shoulders and pug nose.

Semyon's mind started racing. Victor had actually gone over this once, in the classroom at MIT. Sooner or later, he had said, you're going to run into serious heat. Even though the casinos will have no idea what you're doing, they'll see that you're winning, and sooner or later, they'll react. They'll see you raise your bet—and even though no card counter in the world raises his bet that much in the third hand of play—they're going to wonder if something is up. The thing to remember is that you aren't cheating. There's nothing they can do to you. Just get up and leave. Head for the door. Run.

Run.

Semyon tried to rise from his seat, but the hand on his shoulder was too heavy. Not threatening, exactly, not crossing that barrier that would make it an assault. But heavy enough to keep him in his place. Owen hadn't gotten much farther. He was standing, but unable to break through the ring of heavyset guards. Instead, he had started shoving the chips they had won into his jacket pockets.

The pit bosses made it around the blackjack table, and the one with the phone stepped forward.

"Dr. Nussbaum, Dr. Powell. We'd like to have a word with you, if we could."

Semyon was about to respond, but Owen broke in first.

"Well, fucking go right ahead, then. We're all ears."

His voice sounded wild, and Semyon hoped he'd keep his cool. The three techniques weren't illegal, nor was card counting, which, despite their unlikely betting pattern, he guessed was what the casino suspected them of. But losing one's temper and getting violent with a security guard—that was a different matter.

"I think we'd all be more comfortable in one of our back rooms," the pit boss responded calmly.

Owen glanced over at Semyon, and Semyon could see the concern in his eyes. A casino's "back room." Not a pleasant thought. Of course, everyone who played blackjack seriously had heard stories about Vegas back rooms. Everyone had seen the *Casino*s, and *The Godfather*s, but that was a thing of the past, wasn't it? Mobster myths, old-world stories? That sort of thing didn't happen anymore, did it?

Finally, Semyon shrugged. They hadn't done anything wrong. The casino was going to have to let them leave eventually. If they made a big enough scene, they could probably avoid the back room—but they'd certainly be finished at this casino. Maybe if they just went along with the pit boss's request, they'd avoid future problems. In any event, he didn't see what choice they had.

He nodded at Owen, who lowered himself back onto his seat.

"Yeah," Owen said, "that sounds real comfortable."

Semyon relaxed his shoulders, trying to control his nerves. He hoped he was making the right call. Maybe if they started shouting, starting running for the door—no, they didn't want to overreact. The casino just wanted to ask them some questions. Try and figure out how they had just won so much, so fast. Victor's three techniques were a flashy system, this was bound to happen sooner or later. It had just happened sooner.

But as he and Owen were led away from the table by their entourage of six security guards and three pit bosses, Semyon couldn't help thinking back to something Victor had told him once, almost as an aside.

Just remember, they're always watching you. There are cameras everywhere in the casinos. In the gaming area, in the hallways, in the elevators, even in some of the bathrooms. Hell, the only place they don't have cameras is in those infamous back rooms.

Now Semyon and Owen were on their way to that one place in the casino that didn't have cameras. That one place in the casino—where nobody would be watching what happened to them.

CHAPTER 11

A Casino on the Strip
Las Vegas, Nevada

To Semyon's surprise—and relief—he never actually made it to the back room.

He and Owen—and their menacing entourage—had just passed through a pair of heavy wooden security doors at the far end of the gaming area when they were met by two more men, both in their mid- to late forties. One of the men was obviously another casino employee, though judging by his charcoal gray suit he out-ranked the pit bosses, and certainly the security. Semyon guessed from his well-coiffed white hair and crisp clothing that he was either the floor manager or someone even higher up, perhaps a casino supervisor—though he never said anything of the sort. In fact, he didn't even look at Semyon or Owen as he quietly discussed the situation with the pit boss from inside the casino. But it wasn't the man in the charcoal suit that drew Semyon's attention, it was the other man, standing a few feet back, leaning nonchalantly against a stark cinder-block wall.

Thin, not waifish, but stretched out enough to make him look taller than he really was. Angled features, high cheekbones, and nar-row blue eyes, framed from above by wavy locks of blond hair. The hair wasn't long, but it wasn't short either, and it looked like it had been recently styled in an expensive salon. To go along with the hair, the man was wearing an even more expensive designer suit. Victor would certainly have approved; eggshell blue, narrow in the shoulders, cut with double, wide lapels. Beneath the suit jacket, the man's shirt was white, with French cuffs that showed a few inches at the wrists, held together by flashing silver links. His shoes looked foreign, the textured, cured hide of some unknown reptile, mammal, or amphibian, tapered to pitch-black points. In total, he cut a striking figure, lounging against the wall. Like the casino boss, he wasn't looking at Semyon or Owen either, but he wasn't avoiding them. He just seemed—unconcerned.

When the casino's boss finally finished speaking to the pit boss from the floor, he stepped back and had a word with the dapper blond man by the wall. The blond man spoke in low tones, and Semyon thought he caught a hint of a British accent; then the boss nodded and spoke toward the security guards in a clipped voice.

"Take the one with the mouth on 'im to my office. Galen's gonna have a word with his buddy, here."

Before Semyon could say anything, the guards were leading Owen farther down the hallway. The casino boss followed, leaving Semyon alone with the man with the blond hair and the immaculate suit. Semyon was about to protest—he didn't want to be separated from his friend—but decided it probably wouldn't do any good. He guessed what was going on here was a case of good cop, bad cop—but he wasn't really sure which cop he'd drawn.

"Dr. Nussbaum, is it?" the man with the suit said, stepping forward, holding out a thin, manicured hand. "I'm Jack Galen. I'd love a moment of your time. There's a coffee shop just down the Strip which I'm quite fond of. Shall we?"

Semyon stared at the proffered hand. The man's British accent and obvious charm were disarming, but Semyon was still standing in a hallway deep in a casino, and his friend had just been led away, God knew where. Then again, if this Galen fellow was really going to bring him to a coffee shop on the Strip, that meant Semyon would be pretty much free to go whenever he wanted.

So you're the good cop, he thought as he shook the man's hand.

"Sounds great. But if you're planning to break my kneecaps when we get there, I'd like a cup of coffee first."

Galen smiled, his teeth white and straight.

"Personally, I prefer tea before kneecapping, but then, I'm English, aren't I? I've got certain standards to live up to, you know."

With that, he turned and led Semyon back out onto the casino floor.

Ten minutes later, they were sitting next to each other on black wire chairs, overlooking a fairly quiet section of the Strip. They could still barely see the grand entrance to the casino in the distance, but the four-lane highway, crowded with cars heading home from a long night of partying, obscured most of their view. People strolled by on the sidewalk in front of them, most looking as haggard as Semyon suddenly felt. It was amazing how much a card game could take out of you, especially when you were trying to concentrate on a dozen things at once. And Semyon was quickly beginning to realize that the cards weren't the only game going on in this town; Jack Galen was as much a player as Victor and his crew, only his stakes were a little bit different. From what Semyon had gathered, Galen didn't read cards, he read people. And right now, he was trying to get a bead on Semyon.

"See," Galen said, holding his tea in front of him, sloshing the warm liquid back and forth so that it licked at the lip of the ceramic cup. "In Las Vegas, life really is a gamble. Take this tea, for instance. Brewed by a woman with fake breasts and an atrocious dye job, a former stripper who took the money she made shaking her assets at the Crazy Horse Two and opened a coffee shop on the most trafficked sidewalk in the free world. And yet, somehow, it's good. Quite good, actually."

He touched the cup to his lips, then set the tea back on the table. Semyon's own coffee was untouched, still set next to the bill that the waitress had left for them shortly after bringing them their order. Semyon didn't feel much like coffee. He wasn't sure what the thin, well-dressed Englishman wanted from him, but he had half an urge to simply stand up and walk away. Still, he was also genuinely curious because this was something Victor hadn't described, this wasn't part of the game he'd practiced and perfected.

"So you work for the casino," he said bluntly.

Galen ran a finger around the rim of his cup of tea, then touched it to his lips.

"I work for a lot of casinos, Dr. Nussbaum. I'm a consultant, of sorts. I help them figure things out. People, that is. I help them understand people."

His accent was more than disarming; it was like velvet against bare skin.

"And you chose me, and my colleague, to try and understand."

Galen nodded.

"Actually, I just happened to be in the security booth when you and your friend walked in. I'm not sure what drew you to me—or me to you. But I decided, for entertainment's sake, to watch your play. You exhibited an interesting—style."

Semyon didn't like the way Galen hit that last word. He crossed his arms against his chest, trying to appear as composed as possible while feigning a liberal dose of righteous indignation.

"We weren't cheating. We were just getting lucky, that's all. And I expect to get our money back—"

"I didn't say you were cheating," Galen interrupted. "And you'll get your chips—your partner probably still has them in his pockets. I just said you had an interesting *style*. Raising your bet like that on the third hand. The dealer busting like he did—very fortunate thing, that."

Semyon shrugged. Then he looked carefully at the man's narrow blue eyes. The man was smart, obviously. But Semyon was used to smart. If Galen had had any real idea what Semyon and Owen had truly been up to, he would have simply barred them from the casino, added their names to the list that all the casinos kept of unwanted card counters and cheaters. Obviously, Galen didn't know anything about Victor's techniques—how could he? So this was just a fishing expedition. Semyon felt bolstered by the thought. This guy was unnerving, but nothing to be afraid of.

"So you're a private eye," Semyon stated, with certainty. "With Griffin?"

Victor had mentioned the Griffin Investigative Agency a number of times during their training sessions. Griffin was the infamous, notorious private-eye agency that tracked down card counters and cheaters for most of the casinos in Vegas—most of the casinos in the world. Victor had told Semyon that sooner or later, any cardplayer worth his salt ended up in the Griffin face book—a book of Vegas's unwanted, listed in order of dangerousness to the casino's bottom line. It was a mark of pride, actually, because even if Griffin knew you, it didn't mean you couldn't play. It just meant you had to play smarter.

Galen shrugged, not answering Semyon's question. Obviously, he wasn't going to give out any more information than he needed to. Maybe he worked for Griffin, maybe for someone else. Maybe he didn't work for anyone at all.

"You're not a card counter," he finally said, placing his hands flat on the table, inspecting his perfect cuticles. "Of that, I'm pretty sure. And I don't think you're working with anyone in the casino. That dealer you were playing against has been at the casino for years, and he's about as clever as the guys who come through the pit with their wonderful systems to beat roulette."

He looked up from his hands, peering directly at Semyon.

"But you're also not who you say you are. You're not a dentist, and you're not from Cleveland. Your IDs were quite pathetic, actually. Maybe good enough to fool a pit boss on the graveyard shift, but beyond that, pathetic."

Semyon leaned back in his chair. He wasn't sure where his confidence was coming from, but at the moment, he was feeling pretty good. He had avoided the back room, and this guy had nothing on him. He hadn't cheated, he hadn't been counting cards, and he hadn't done anything wrong. He had just won, that was all.

"Maybe I'm just a high roller who values his privacy."

Galen cocked his head to the side, as if considering the thought.

"Maybe. Certainly, we've got our fair share of those. If so, then I've made a frightening mistake, haven't I?"

Then he shook his head. His wavy blond locks barely moved, lacquered as they were with thick coats of coconut-scented mousse.

"But somehow, I don't think so."

Suddenly Galen rose from his chair. The impromptu meeting was over—just like that. Semyon was glad; he didn't want to be around the Englishman any longer than he had to. The man wasn't frightening, exactly, he was just—unnerving.

As if on cue, Galen leaned forward over the table, his face only inches from Semyon's.

"I don't know anything about you, 'Dr. Nussbaum.' You're a blank slate. But for some reason, I have a very good feeling I'll be seeing you again."

He stepped back from the table. Semyon felt his stomach tightening, but he fought the feeling back. He reminded himself: The cooler he appeared, the less he had to fear from this bizarre "consultant."

"So you're the good cop, right, Mr. Galen?"

Galen smiled as he suddenly tossed a perfectly aimed, perfectly folded ten-dollar bill on top of the check, with a flick of his manicured fingers.

"In Las Vegas," he said, "there aren't any good cops."

Three hundred yards away, in a windowless, cameraless, cinderblock room in the basement of the casino, Owen Keller licked blood off of his lower lip as he struggled against a pair of cold metal handcuffs. He was seated on a steel chair in the center of the otherwise empty room. His arms were pulled back behind his back, fastened by the cuffs around his wrists to the base of the chair, and the blood was from a fresh cut just above his mouth. The cut hurt, but the pain wasn't important, at the moment. Owen had a feeling he had a lot more to worry about than a scratch above his lip.

Dazed, scared, his adrenaline blasting at full throttle, he couldn't begin to guess how long he'd been handcuffed to the chair. He wasn't sure what time it was because like the rest of the casino, the room didn't have any clocks. Just a bare lightbulb hanging from the ceiling, the steel chair, and those cinder-blocks, so close and confining that Owen could feel his own hurried breaths blasting back at his face.

Fuck, this wasn't good.

He had no idea what they were going to do to him. The security guards hadn't said a word as they'd led him to the room. Then, as they'd pushed him inside, he'd overheard two of the fuckers talking in the hallway, in purposefully loud voices. Something about how "the guard he had hit wasn't going to make it." Bullshit, of course, Owen hadn't hit anyone. The truth was, the only one who had done any hitting was the Mongoloid security guard who had cuffed him to the chair. Not a punch, really, just the back of a hand against Owen's face. But from the way the guards outside had been talking, it sounded like they might be planning on setting him up. Maybe calling the real cops, having him arrested for an imagined assault.

Obviously, they wanted to play rough. Well, Owen had played rough before. This wasn't the first time he'd been in handcuffs.

But deep down, despite his bravado, he really was scared. Because the truth was, this wasn't a police station, and these weren't cops. This was the basement of a casino. These were underpaid, overeager security guards, the lowest on the casino totem pole. And this place was a back room. Anyone who'd ever been to Vegas instinctively knew—the back room was a place you didn't want to end up.

Especially not handcuffed to a chair, with over one hundred and fifty thousand dollars of the casino's money jammed into your pockets.

CHAPTER 12

Caesars Palace
Las Vegas, Nevada

PRESENT DAY

The music was so loud that the chandeliers were actually moving with the beat. A throbbing, pulsing sound that I could feel in my chest, each burst of bass hitting me like an overcharged defibrillator blast. It didn't help that the dance floor was separated from our red-on-red VIP cocoon by a doorway and a throbbing mass of beautiful people; it felt like we were right in the middle of the sound storm, swept up by the sweaty current of scantily clad dance-club patrons.

"This place rocks, doesn't it? Ever seen anything like it?"

The young man with the dyed black hair and tattooed arms was shouting at me across a miniature glass city of vodka bottles, mixers, and ice buckets, but still I could barely make out his words. It didn't help that he had a thick, but indistinguishable European accent, or that the piercing in his bottom lip made him slur most of his consonants.

"It's pretty impressive," I shouted back.

I wasn't just making conversation. The club really was quite a scene. A massive, thirty-six-thousand-foot venue, with two floors and three distinct settings for late-night fun. The first floor had been done up in various shades of white, from the curtains that lined the walls, to the huge dance floor, to the raised VIP seating area. Instead of chairs, the first floor had beds, where incredibly expensive bottle service was available.

Here, in the separated VIP area, we were drowning in a sea of red. Plush red curtains, red upholstered walls, red lighting, red every-thing. Of course, it was called the Red Room, but I wasn't going to fault the club for it; sometimes, it was good to be literal.

"If you take the elevator up to the top," the tattooed man contin-ued, "there's another dance floor. Oval, really cool. Open year-round. Panoramic views of the city."

He was bobbing his head now, in beat with the music. I watched him from my perch on the edge of a red leather couch and contem-plated pouring myself another drink. I didn't think it would help with the noise, but maybe it would get me more into the spirit of the place. When my interview subject had suggested we meet here, at this new club in Caesars Palace—at two in the morning, of course—I'd assumed he was just trying to pick someplace neutral, where it would be hard to get his voice on tape. I'd run into this sort of thing many times before, people who'd assume that I'd have a tape recorder and a spy camera hidden on me, like some sort of private eye. I wasn't that sort of writer, but that was difficult to explain, so usually I just went along with the game. But in this case, I didn't think my tattooed and pierced subject had suggested Pure because he was afraid I was going to get him on tape. Alex Peyton had sug-gested Pure because this was where he liked to spend his Saturday nights.

Although it was hard to believe, Peyton was the one who had once worn wires and carried a miniature camera. Up until about three years ago, he had been employed by a local security "consult-ing" firm that worked for many of the casinos in town. Although he was too young to have ever worked with Jack Galen, he had walked the same beat for more than five years and knew as much about the cat-and-mouse game the casinos played with high rollers like Se-myon Dukach and Victor Cassius as anyone around. He'd done work for the various PI firms in town and had consulted for almost all the casinos: especially Caesars, where we were sitting now. He knew all about the infamous back rooms—Caesars had its own version of a room very much like the one where Owen had found himself handcuffed to a steel chair.

"You know, you don't look like a PI," I said as my eyes wandered across the VIP area. I saw a few faces I recognized from *US* magazine and *People*, the usual club set that frequented places like this. On Saturday nights, Vegas had become an extension of Hollywood, a place for the pretty people to make the scene and be seen. For many of them, it wasn't about the gambling, it was about the nightlife, the megaclubs that were rapidly overshadowing the megaresorts. Places like Pure were a reason to come to Vegas now, which in many ways was a return to the concepts that had first put the neon city on the map. Bugsy Siegel's concept of a "carpet joint," reimagined for a new kind of Hollywood jet set. The Paris Hilton set.

"Ha," Peyton responded, pounding his hands on his knees with the music. "Ninety percent of PI work is high-tech now. Computers, cameras, software. You're trying to find information about people on the electronic highway, not the asphalt one outside. Do I look like a hacker?"

I laughed. Actually, he did. I shouldn't have been surprised by his appearance. Semyon was the one who had given me his contact information, and they had met at a computer convention. As far as I knew, Peyton didn't know about Semyon's blackjack past, just about his facility with computers. Though they were on opposite sides, they had more in common than not.

"Weren't a lot of the guys you were chasing down for the casinos really just hacker types as well?" I asked as a pair of blond stripper types passed close to our table, leaving behind a thick wake of cherry-scented perfume.

Peyton shrugged. "You mean the card counters, the math guys with the systems, right? Yeah, some of them were hackers, but some were outright cheaters. There's a difference."

This was something I'd been thinking a lot about. That gray area between using math and skill to beat a game that wasn't supposed to be beatable, and outright cheating.

"Where do you make the distinction?"

"Well, remember, it wasn't my job to make the distinction. My job was to find out as much as I could about the people who won too much—"

"Too much?" I interrupted.

"Who won more than they were expected to win." Peyton grinned. I knew why he was grinning; it was one of the many secrets of the neon city. The truth was, the casinos looked into anyone who won consistently in Las Vegas. High rollers, weekend players, tourists, anyone who won. You were supposed—expected—to lose. Because if you won, consistently, you were an aberration that had to be explained. And eventually, dealt with.

"So my job was to check out these 'persons of interest.' Not the scam artists and the blatant cheaters—those guys the casino's security could handle. But the others, the ones who were winning but the casinos couldn't figure out why. Ninety percent of the time, they really were just getting lucky. And ninety-nine percent of the rest were card counters. College kids from MIT or Caltech, places like that, who could find advantages in blackjack. No big deal. I gave their pictures and their names to the casinos, they ended up in the Griffin book, and they eventually got barred."

I nodded. Because of my previous book, I knew a lot of those guys personally, the who's who of the card-counting world. For the most part, most of them didn't harbor any ill will toward the techies like Peyton who helped hunt them down. It really was just another element in the "hack."

"And the other one percent?" I asked.

"Those were the ones that the casinos really went after. Because most of them really were cheating the system. Not beating it, cheating it. We just didn't know how."

I leaned back on the couch. I knew that Semyon did not consider what he had done cheating. He was manipulating the cards, but not by doing anything outside of the rules of the game. Just by hitting and not hitting his hands. By using the cut card that the casino allowed him to use. By choosing dealers who shuffled in a certain way, who handled the cards in a certain way.

Was it wrong to see the back card in a rolled-over shoe? Was it wrong to use that knowledge to win? In a friendly game of poker, if I saw my neighbor's cards and played accordingly, it might be considered cheating. But in a casino, where the odds were stacked against you from the start, where you simply took advantage of your opportunities, was it wrong to look for that back card?

"See," Peyton said, thoughtfully touching the metal hoop that bisected his lower lip, "it's really not the winning that matters. It's the fact that the casinos can't figure out how they're doing it. That's what scares them. Because if *you* can do it, maybe you can show others how to do it. And then, maybe, the casinos are going to have a problem."

I felt a chill down my spine as I thought about the book I was in the process of writing. Because that's exactly what Semyon and I were going to do. Show others how to do what Semyon had done. Unlike card counting—which the casinos already knew about because there were hundreds of books and TV shows and documentaries on the subject—Victor's three techniques were something new. Incredibly powerful, and new. How was Vegas going to react?

I thought about that back room downstairs somewhere in the casino, with its own version of the cinder-block walls and the steel chair.

"Didn't you ever worry that the information you gave the casinos about people 'cheating' would lead to—"

"Violence?" Peyton interrupted, his grin back. "Fingers getting broken and kneecaps getting shot off? Ha. This isn't Hollywood, though sometimes it feels like it. The thing is, as casinos go, Vegas is about the safest place in the world. You used to have the mob, but that's all gone now. It's all major corporations, five-star restaurants, hell, we're sitting in a massive nightclub in a fake Rome. This is Disney World."

"But there are back rooms."

He shrugged. "Sure. I've seen 'em. I've been in 'em. No big deal. It's just an intimidation technique. They put you in there for a few hours, let you soak in your own sweat. The casinos know that you've seen all those scary movies, they know you're expecting a big dude with a hammer to come out and smash your knuckles to pulp. But they're not really going to do anything like that. It's not worth the lawsuits."

He began drumming his knees again as the music from the nightclub seemed to get even louder.

"But Vegas isn't the only place with casinos, and a lot of these one percent dudes, the ones with the systems we can't figure out—they react to the heat in Vegas by going elsewhere. And that's where it gets dangerous. Because in some places, casinos are mom-and-pop operations. In some places, the casino is literally the lifeblood of the town."

He paused, then his face got a little bit darker.

"And overseas, in the Caribbean and Europe, you've got real mobsters. Italian, Russian, Chinese. These people take things a bit more seriously. You find a way to beat the system over there, you're stealing from the mob."

He looked at me, and I had the sudden feeling that he knew more about Semyon's story than either of them had let on.

"You know what happens when you steal from the mob?"

CHAPTER 13

Boston, Massachusetts

emyon wasn't afraid of heights, exactly; he had no problem with airplanes, had never fallen out of a tree, had never had any trouble with elevators or balconies or ladders. He'd even tried rock climbing once, on the advice of an ex-girlfriend, but even after two days clinging to some craggy rock in New Hampshire, as he worked his way to a summit littered with empty Gatorade bottles and discarded PowerBar wrappers, he never quite got the point of it. Climbing equipment was expensive, anyway, and for the most part, Semyon had always wanted hobbies that *made* him money, not cost him money.

But his thing with heights wasn't about mountains, airplanes, or balconies; it had to do with very tall buildings, specifically, skyscrapers with pretty little restaurants on top. Whenever he met anyone for a drink at one of those postcard-perfect, perched-in-the-stratosphere, glass-walled tourist traps, he also had the same urge. To take his chair, toss it through the double-plated glass, and leap out into that postcard. It wasn't about glorified self-destruction or suicidal vandalism; it was just the sense that everything was just too determinedly perfect. A city wasn't a city from a thousand feet in the air; it was a picture, unreal. Semyon didn't know why, but places like that always made him feel incomplete.

Maybe one day, he'd try skydiving and solve that puzzle for himself. At the moment, his face against cold, double-paned glass, gazing out a spectacular picture window at a spectacular view, he thought that the Top of the Hub would make a great launchpad for his first skydiving experiment.

The Top of the Hub was one of those places that you went to only when company was in town. Perched atop the Prudential Building, Boston's second tallest skyscraper—and the city's answer, albeit a provincial one, to New York's Empire State Building—the Top of the Hub was an elegant jazz bar/restaurant with stunning 360-degree views of Boston. The decor of the place was New York chic, with plenty of muted vinyls and soft leather; the jazz band stayed in a corner, as far from the diners as was humanly possible, and the place's bread and butter were tourists from all over the globe who'd somehow gotten the idea that Boston had more to offer in terms of nightlife than frothy college bars, underground rock dives, and youth-oriented discos. The Top of the Hub was one of the very few places in town where adults actually went out for drinks, ones replete with funky names and served in tall art deco glasses, and more important, it was one of the few places in town where it was unlikely you'd run into a pack of drunk fraternity pledges or freshman girls out for their very first beer.

In all, a good place to meet friends or relatives from out of town. Which was why Semyon had chosen it, why he had been standing by the window, gazing out over the Back Bay, at the snaking, serpentine Charles River and the pretty, parallel rows of nineteenth-century town houses, while he contemplated how hard he'd have to throw one of the vinyl chairs at the double-paned glass to get it through to the other side.

"Semyon. Hey, look at you, boychik. You got a haircut."

Semyon turned just in time to see the huge man barreling toward him, thick, meaty arms out at both sides like decidedly nonaerodynamic airplane wings.

"Dad. You made it early."

Semyon's father grabbed him in a strong bear hug, nearly lifting him off his feet. The few tourists nearby looked and smiled, probably commenting to one another on the huge physical differences between the two relations; where Semyon was thin and angular, his father was wide and rounded. A big man—no, a huge man—though now that he was getting on in years, he was beginning to hunch forward a little, belying his true size. Growing up, Semyon had always known that his father was physically intimidating to those who didn't know him; but in reality, David Dukach was one of the most gentle—if giant—Russians who'd ever grown up on the mean streets of Moscow.

After the bear hug finally ended—and Semyon found the floor with his feet—they moved to a pair of vinyl couches right up against the picture window. They were far enough from the jazz band to be able to hear each other speak, but not so far that they'd have to fill any silences that were bound to spring up. Semyon had always had a good relationship with his father, but they'd never really had that much in common, in terms of lifestyle. Their common ground had always been math and science. David was a brilliant man, an engineer, but the hard times he'd had bringing his family out of Russia and bringing them up in some of the poorest neighborhoods in America had not allowed him much time to develop modern hobbies or modern tastes.

"I managed to talk my way onto an earlier flight," his father responded, after a waitress who looked like she was still in high school took their drink order—vodka martini for Semyon and straight vodka for his father. "This flying, these airports, those little seats—it's all exhausting, isn't it?"

Semyon nodded, but the truth was, he was pretty sure his father had flown maybe twice in the past year. David didn't leave Houston often, partly because he couldn't really afford to and partly because he really didn't want to. He had his routines, his simple life. His Russian friends whom he drank with on weekends. His work—computers, circuit boards, electronics. And his television, mostly news programs, which he watched voraciously. He'd never been much of a vacation person and had only visited Semyon when his wife—Semyon's mother—prodded him out of the house. Usually because she had something she wanted to tell Semyon, something in his behavior she wanted to correct. When he was a kid, it was always the clichéd "wait till your father gets home." Now, as an adult, Semyon had gotten the phone call that his father was on his way to the airport. The meaning was still the same.

"Your mother wants me to tell you to stop with this blackjack thing," David said, bare seconds after the waitress placed their drinks on the table in front of them.

And there it was. Semyon had told his father all about Victor, the three techniques, and his first weekend in Las Vegas. Semyon had always told his father the things that were going on in his life—good and bad. Actually, sometimes he liked to weight his conversation with the bad more than the good. He'd always had a bit of a rebellious streak when it came to his parents, and even as a kid, he'd always taken the opportunity to jab at his father when the opportunity presented itself. In fact, many of the hobbies he'd had over the years were directly the result of his urge to piss off his dad. His early obsession with video games, which his father had always seen as a waste of time. His comic-book collection, which his father had likened to playing with kindergarten coloring books. And of course, cards.

"I know you believe it really works," his father said, taking a sip of his vodka. "These techniques. This sequencing of cards. But to me, it just sounds so, well, impossible. And your mother doesn't like the idea of you going to Vegas. She doesn't like the idea of you gambling."

Semyon rolled his eyes. He'd tried his best to explain the system to his father—that it wasn't, in fact, gambling per se. They knew the mathematical outcome before they began. It was amazing to him that his father just couldn't get it. David Dukach was a man who took apart microwave ovens and put them back together just to see how they were made. He'd built a computer from scratch in the basement of their house in Newark. And he was having trouble believing that you could follow a group of cards through a six deck shuffle? No wonder the casinos were never going to figure out what Victor had come up with. No wonder they were going to make millions.

"It works, Dad. Better than you can imagine."

David had a skeptical look on his face. His lips were turned down at the corners, almost lost in the tufts of reddish fur on his jaw.

"Okay, let's say it works. Is it dangerous?"

Semyon quickly shook his head—but deep down, he couldn't help but think back to Owen and what had happened to him in that back room. When they'd finally been reunited, back at the Mirage suite, he'd been shocked to see the blood on Owen's lip, and the bruises from the handcuffs on his wrists. Thankfully, the thugs at the casino had not carried out their implied threat of setting him up for some fake assault charge; after three hours stewing in that cinderblock back room, Owen had finally been released. Without a word, without apology or explanation, the security guards had simply led him out a side door and deposited him back on the Strip. No police involvement, no trespass speech, no nothing. They hadn't even asked to see the chips in his pockets.

Victor had quickly gathered the team together and then had explained that the entire episode was all an effort at intimidation. The casino had just wanted to scare the hell out of them. They had chosen Owen because he had mouthed off to them, and also because he looked a little more rough-and-tumble than Semyon. The way Victor had said the last part, Semyon thought that he was implying something more—but he didn't elaborate, and Owen didn't complain.

Even without that, Semyon wasn't especially pleased with the way Victor had brushed off what had happened to Owen. Semyon knew that Victor had never really liked Owen, hadn't really wanted him on the team. But still, he should have had more sympathy for what Owen had been through. Handcuffed to a chair? Hit in the face? Semyon would have been terrified if that had happened to him. All he'd gotten was a cup of coffee and a bizarre—if slightly threatening—bit of conversation with the dapper "consultant" Jack Galen.

Still, after going over it again and again in his head, Semyon had decided that the entire incident had been more frightening than anything else. Overall, the weekend had been a stunning success. They'd made more than two hundred thousand dollars, had stayed in an incredible high-roller suite, and he'd successfully used all three techniques. What's more, he had gotten to spend time with Allie, which felt almost as good as winning a few big hands of blackjack. He wasn't going to let the one incident dissuade him from playing the game he'd discovered. If he'd been the one handcuffed to the chair, he might have had a bit more hesitation on continuing the path he'd started—but the conversation with Galen had actually inspired him. He felt challenged by the man, and he'd always had a competitive streak. Especially when it came to games that involved his intelligence.

"It's not dangerous," he finally responded. "The casinos don't have a clue what we're doing. We're going to make a lot of money, Dad."

His father perked up in his seat, and Semyon smiled inwardly. That was the weak spot, the chink in his dad's armor. Money. David Dukach had never had any money. His family had gotten by on what he could earn as a low-level engineer, and what he could get on the side for fixing the Orthodox Jewish neighbors' microwaves, radios, and TV sets. No matter how much he disliked Semyon going to Vegas to play cards, the idea that you could win a lot of money with the three techniques was a powerfully compelling argument.

But that argument was just the beginning because like any good card schemer, Semyon had an ace up his sleeve. Or more accurately, in his pocket.

He reached into his pants and pulled out two slips of colored cardboard. He handed them across to his father. His father turned them over, looking at the glossy pictures on both sides. They were tickets. Incredibly expensive and very hard to get.

"The Holyfield fight," Semyon said. "Front row, center, three weeks from now, at the MGM Grand—in Las Vegas. And the tickets come with free airfare."

David's eyes went wide. He held the tickets like they were encrusted with diamonds.

"Where did you get these?"

Semyon smiled. His father was a huge boxing fan. One of his earliest memories was watching his father hitting the side of their crappy black-and-white TV, trying to get better reception so that he could see Muhammad Ali's face as he beat up Sonny Liston.

"I'm a high roller, Dad. This is the kind of thing the casinos give you when you're a high roller."

"But what do you have to do to get these tickets?"

Semyon's smile turned into a true grin.

"I just have to play more blackjack."

David opened his mouth to say something, then closed it again. Instead, he took a sip of his vodka. Then he cleared his throat.

"When's your next trip. With this Victor person, and your team."

Semyon noticed the slight change in his father's voice. Before, he'd talked about the blackjack scheme as if it was something dirty, like cheating. But now the way he said "team" sounded more like the description of a sporting excursion. *It was amazing, how well bribery could work.*

"Next weekend. We're going to Atlantic City."

It had been Victor's idea to give Vegas a break for a couple of weeks, after the handcuff incident. He wasn't worried about the heat; he had agreed with Semyon's assessment that Jack Galen was just fishing, that he had no real information to go on. But when the heat got hot, it was always good to take a little break, let things cool down a bit. There were plenty of casinos, all over the country. All over the world, actually, and Victor had intimated a few times that he was planning some real blackjack trips in the near future. But Atlantic City was a great place to start. It wasn't as glitzy as Vegas, but there were places there that could handle their action, and that's what really mattered.

Semyon's father rubbed his free hand over his scruffy jaw, mulling it over. Then, finally, he shrugged his thick shoulders.

"There's a nice Russian bakery on the boardwalk," he said, winking conspiratorially at Semyon. "Right near Harrah's. I'll tell your mother you're going for the bialys."

Then he held up the Holyfield tickets and grinned.

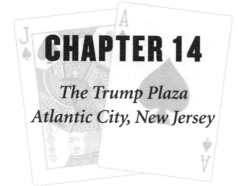

CHAPTER 14

The Trump Plaza
Atlantic City, New Jersey

I think I could get used to this."

Semyon clinked the edge of his wineglass against the smoke black window of the limousine as the first bright lights of the Boardwalk began to flash by. The limo was sleek and black, snaking through the traffic as it wound its way toward the only true gambling mecca on the Atlantic Ocean. The gray interior of the car smelled of leather and sported a TV, a wet bar, even a small refrigerator filled with snacks. The Plexiglas divider that separated the passenger cabin from the uniformed driver was polished so severely that reflections flashed with the clarity of images on a movie screen: Allie, in one corner of the backseat, tight leather pants and cutoff baby T-shirt, elbow crooked against an armrest, hand in her deliberately messy blond hair. Owen next to her, back in his black leather jacket, holding his own glass of wine with both hands while he took a long swig. Victor, in profile, hugging a wine bottle against his silvery blue suit. And Semyon, in jeans and a black Armani shirt, watching the lights of the city bounce from the window to the curves of his crystal wineglass.

The wine was red and strong and expensive—more than twenty-five hundred dollars a bottle, Victor had explained as he'd first uncorked the '82 Château-Margaux. He'd further explained that only the '82 and the '66 were worth drinking, though the '66 was well past its prime—so he'd made sure that they'd have the limo stocked with the '82. Sure enough, two bottles had been waiting for them in a bronze ice bucket when the driver had shown them to the car.

If the man in the lime green uniform with matching hat had known that the four passengers he was picking up at Newark International Airport had actually flown from Boston on a cut-rate junket—that they'd actually cashed in coupons at the gate to save fifty bucks on their already cheap seats—he'd probably have tossed them out at the first stoplight. But all the driver knew was that they were VIP guests of the Trump Plaza, which meant they were high rollers, whales, precious cargo. The type of players who drank twenty-five-hundred-dollar bottles of wine during the forty-minute ride toward the oceanfront casino.

"It looks like we're almost there," Owen commented as the lights grew brighter outside the tinted windows. It was the first time, since the airport, that they'd seen anything other than overpasses, truck-stop diners, gas stations, and chemical plants. Now there was a glow from a small constellation of bright lights in the near distance and the first signs of the famous Boardwalk that ran three miles down the shoreline. The lights seemed huddled together, an oasis from the asphalt maze of highways. "Doesn't look like much."

Victor waved the bottle of wine toward Owen, but responded to the rest of them. Semyon had noticed that Victor rarely addressed Owen directly.

"Atlantic City's built on a grid, like the Back Bay. But it's got very little in common with Boston."

He was in full sensei mode, channeling the genial MIT instructional style they were all overly familiar with.

"From the east, there's the sea, the beach, and the Boardwalk. The Boardwalk, up ahead, that's where it's as bright as Vegas—but razor fucking thin, one tiny strip of light, a bare handful of casinos clumped together."

He pointed out the window; a kitschy scene of Americana, a boardwalk diner with vinyl stools and an outdoor seating area. Seven P.M., and already a crowd was milling about the street corner, some gathered by the diner, others gawking at a street performer, a woman with no arms playing a pair of bongo drums with her chin.

"See, it starts here, and goes about a mile down the ocean. On the other side of the casinos, that's Pacific Avenue. Still glitzy, but a different sort of glitz. Strip clubs and pawnshops and check-cashing dives. The strip clubs aren't bad, the girls are pretty and they wear G-strings; but they make up for it by happily whacking you off under the counter."

Semyon glanced at Allie, but if she was offended or even bothered by Victor's remark, she didn't show it. Owen was still turned toward the window, but Semyon could see the edge of his patented smirk. If Carla and Misty had been there, they probably would have responded with some sort of crack about G-strings and hand jobs, but they had caught a later flight with Jake; seven of them in one limo would have been crowded, and less believable. Anyway, Jake had played enough in AC to warrant a limo of his own.

"After Pacific," Victor continued, "there's Atlantic Avenue, but you don't ever want to go to Atlantic, because there's nothing for you on Atlantic besides drug busts and drive-bys, and the inner-city jungle in its full, glorious decay."

"Sounds charming," Semyon said, finishing his wine.

"Oh yes, very charming. Actually, the truth is, Atlantic City sucks. Not just because the city is mostly a dump. But it's actually a really bad place to play cards."

"Why is that?" Allie asked.

"You can blame Ken Uston for it," Victor said, sighing. Semyon perked up at the mention of the infamous big player, whose books had been responsible for his own blackjack knowledge. "About ten years ago, Uston sued after getting kicked out of one of the casinos. Took the case all the way to the New Jersey Supreme Court—and won."

Even Owen looked up at that one. It was hard to believe that a card counter could win a lawsuit against a casino. Then again, this wasn't Vegas.

"New Jersey isn't Nevada," Victor said, echoing Semyon's thoughts. "The casinos don't run the state, just this little bit of beachfront property. So Uston won, and the casinos here can't throw you out for card counting. Instead, if they think you're a counter, they just fuck with the game. Make the table maximums one dollar. Shuffle up every hand. Pretty much make it impossible to play. So it's a crappy place for card counters."

"Good thing we're not card counters," Allie commented, winking. "Just your run-of-the-mill high rollers."

Victor smiled. "Well, actually AC's not that great a place for high rollers either. They make this big show of trying to attract rich bankers from Manhattan, but the truth is, most of the casinos cater to the lowest common denominator, the welfare gambling addicts and lower-middle-class dreamers who come for the fifty-cent slots and cheap buffets."

Semyon looked around the posh limo, at the leather seats and smoked-glass windows.

"Looks to me like there's at least one casino that can handle our kind of action."

"And that's why I love you." Victor grinned. "You're so fucking observant, Nikolai."

Welcome to the Trump Plaza, Mr. Nogov. It's a real pleasure to meet you."

Semyon looked up from his cards just in time to see the short, squat man with the thinning red hair move across the velvet-carpeted high-stakes pit and sidle up close to his third-base position at the blackjack felt. The man had a wide smile on his face and was followed by a cocktail waitress carrying a tray of four champagne glasses.

"I'm Karl Dover, your host here at the Plaza. Kent Tucker spoke very highly of your play at the Mirage, and we're just happy you chose our casino for your trip to Atlantic City."

Semyon did not need to fake his own smile as he shook the man's hand. The dealer had paused middeal as Dover had approached the table, and Semyon was sitting on two nineteens against the dealer's six. More important, he had two bets of ten thousand dollars each on the nineteens—and Owen, to his right, also had twenty grand out, as did Allie at the first-base seat. And most important of all, Semyon knew, with the certainty of a mathematician looking over a simple problem of arithmetic, that the next card in the deck was a king of spades. That meant that he and his two teammates had sixty thousand dollars on the table with an expected return of 30 percent, per hand.

In the little more than the hour they had been at the Trump Plaza, this was the fourth time they had managed to run the third technique. They'd already busted the young, pretty, and small-handed female dealer twice, for more than seventy thousand dollars in profit. They'd also cut to an ace once—the first technique—for a fifteen-thousand-dollar blackjack. Altogether, they were up more than fifty grand—and that was just Allie, Semyon, and Owen. God only knew how much Victor, Jake, Misty, and Carla were earning running the second technique on the other side of the fairly crowded blackjack pit. Victor had been right to choose the Plaza because indeed, the bustling, garish casino, bristling at every angle with the real-estate mogul's gold-encrusted signature, could handle their action. And then some.

"Thanks very much," Semyon responded, thickening his Russian accent as the cocktail waitress deftly placed the champagne glasses on the table, careful not to disturb the towers of chips, each with Trump's name plastered across its face, in front of each seat. "Mr. Tucker told me that your casino would take very good care of me and my associates."

Dover glanced at Owen and Allie, then quickly brought his eyes back to Semyon. Semyon could only guess what Tucker had told the East Coast casino host about Nikolai Nogov; surely, Dover would make his own assumptions about Semyon's leather-clad colleague and the blonde in the baby T-shirt. Semyon smiled inwardly at the thought that he was rapidly making a name for himself among the casino hosts that catered to the high-rolling set. To the hosts, people like Nikolai Nogov were currency, names that could be traded like commodities because of the kind of action they brought to a casino. It had been Victor's idea for Semyon to call Kent from a pay phone before their trip to Atlantic City; he wanted to build Semyon's name as fast as possible, to lessen the chances that a situation like the one that had occurred in Vegas would happen again. Victor's thinking went that Galen would not have found Semyon and Owen's play suspicious if they'd been better known by the casino beforehand. Having met Galen, looked into those unnerving eyes, Semyon wasn't so sure that was true—but he did like the idea of becoming somewhat of a high-stakes legend. It certainly couldn't hurt his act.

"We'll do everything in our power to make your stay here as enjoyable as possible," the plump host said, his obsequiousness bordering on the obsessive. He ran his eyes over the massive bets Semyon and his team had out on the table, and his face lit up even more. Then he leaned close to Semyon and gestured toward the back of the blackjack pit.

"And when you're finished playing, I'd love to introduce you to your real host for the evening."

Semyon followed Dover's gesture to two men in dark suits standing by a mahogany bar, deep in conversation. The man on the right, gray-haired and slightly stooped in the shoulders, was unfamiliar. But the other man, with his perfect storm of wavy auburn hair, pursed lips, and arrogant stance, was easy to recognize. Donald Trump, the real-estate billionaire in the flesh.

Semyon paused, staring. He saw Owen turn to see what he was looking at, then Allie as well. Semyon slowly digested Dover's offer. Sure, Semyon would have loved to walk right up to Donald Trump and shake his hand. Then he'd love to have told him how he'd taken Trump's casino for tens of thouands of dollars—would continue taking it all weekend long. How he was going to beat the self-aggrandizing bastard at his own game.

Semyon laughed out loud, then shook his head, turning back to Dover.

"That won't be necessary. I think we are very different people, Mr. Trump and myself. I'm just happy to play cards and enjoy the company of my good friends."

He clapped Owen on the back and smiled at Allie, who blew him a mock kiss. Dover raised his eyebrows, then shrugged, nodding at the dealer to continue play. Semyon barely watched as the dealer turned over his down card, a ten, for a hard sixteen. Then he barely breathed as the dealer reached for his draw card—and pulled the predicted king of spades from the deck.

"Dealer busts," the dealer called out, her voice cracking slightly.

"Pay the table!" Owen shouted, and Allie squealed with delight.

Semyon raised a fist in celebration, then gave Dover a thumbs-up. As Dover moved away from the table, Semyon looked past him again, toward Trump and his associate. The billionaire didn't seem to notice the commotion coming from Semyon's table. Truth was, he probably would not have cared about a sixty-thousand-dollar payout. To him, it was pocket change. But if he'd known that it was the tip of the iceberg—that Semyon and his team could continue winning like this, almost at will, all night long—he might have paid more attention.

If he'd known the truth, maybe he'd have been staring at the table like the two men in ill-fitting brown suits who were standing at the other end of the mahogany bar, closer to the high-stakes pit.

Semyon squinted his eyes, focusing on the two men. He didn't think he'd ever seen them before. Both were young, maybe in their early twenties, and looked rough around the edges. One was tall, with wiry limbs and a narrow, scruffy face. The other looked Latino, with slicked-back dark hair and hard, colorless lips. Their brown suits looked cheap, and neither of the men looked like he worked for the casino. Probably they were just two gawkers, getting their kicks watching high rollers playing with bigger money than either of them had ever seen before. Or maybe they were just admiring Allie from a distance. Still, Semyon filed their faces away in the back of his mind.

Then he turned back to the table, where the dealer was scooping up the cards, getting ready to shuffle again. Quickly, the two men in the brown suits faded from Semyon's thoughts as he watched the young dealer working the cards with her small hands. He knew that once again, from her first-base seat Allie was going to try to read the back card when she rolled the stack over. If it was a ten, jack, king, or queen, she'd give Semyon and Owen the proper signal—by folding her arms against her chest—and they'd run the third technique again. An ace, and she'd touch her nose, signaling that they'd be running the first technique. Anything else, they'd just play it through. No matter what happened, no matter how she signaled, they were going to take more of Mr. Trump's money.

Semyon looked past Owen and caught Allie's eye. He winked at her, and her full, bright red lips turned up at the corners, parting slightly to show a flicker of white teeth.

Christ, she had a beautiful smile.

CHAPTER 15

The Trump Plaza
Atlantic City, New Jersey

e've got to stop meeting like this."

Semyon grimaced as he lowered himself to the tiled floor, keeping as much distance between himself and the porcelain toilet bowl as was humanly possible. He'd been spending entirely too much time in bathroom stalls since he'd started playing blackjack. Tonight, alone, he'd visited this particular stall in the back corner of the men's room in the Trump Plaza six times. He'd had to keep ordering glasses of water to go along with his champagne, just to keep up appearances. Better that Nikolai Nogov got the reputation of an overhydrater, a Russian arms dealer with a bladder problem.

He let one knee touch the tiles as Victor's tan hands suddenly appeared, fingers outstretched, from beneath the wall that separated Semyon's stall from the one next door.

"This sure is a glamorous life we've chosen," Semyon grumbled as he began pulling stacks of hundred-dollar bills out of his pockets. Victor took the stacks, sliding them back into his own stall, then reaching out for more.

They had been doing this at intervals all night long. Semyon in the corner stall, on his knees. Victor in the stall next to him, also on the floor, reaching out with hungry hands.

"You failed to mention this part of the gig at my initiation," Semyon commented as their exchange crossed the hundred-thousand-dollar mark.

Despite his complaining, Semyon knew that the bathroom was the safest place to make the cash transfer. There were no cameras in the bathrooms, and definitely no way anyone could see what they were doing from outside the stalls. Since they were hitting the Trump Plaza as two different groups—not as strangers, since they'd come in the same limo, but as high rollers who'd met in Vegas and were re-unioning in Atlantic City—it would look strange for them to be sharing cash on the casino floor.

At the moment, Semyon was transferring most of his stake to Victor, since he and Owen were going to be taking a short break from the action to check out the Boardwalk, maybe do some recon-naissance of the other nearby casinos. They'd already won enough at the Plaza to generate a fair amount of heat—if anyone upstairs had decided to get suspicious—and a few hours downtime would do them good. Semyon was still wary after the incident in Vegas with Jack Galen, and Owen was, understandably, even more paranoid. Owen had suggested the break, and both Semyon and Allie had agreed. Allie had wanted to get a nap in before their next shift of play, so the decision certainly didn't need to go to a vote. Besides, one hundred and thirty thousand dollars in profits was already a pretty good take for a Friday night.

"Don't stray too far," Victor said as his hands withdrew with the final few stacks of money. Semyon kept only a few hundred dollars in his jeans pockets, since Victor was going to continue playing large action at the Plaza and could use the bigger stake. "Remember what I told you about Atlantic Avenue and the rest of this city. It's really fucking ugly out there."

"Just heading down the Boardwalk," Semyon curtly responded. Although he respected Victor's opinions, he really didn't need the fa-thering. He had grown up in neighborhoods as bad, if not worse, than AC. "Maybe check out the Tropicana, see if it's good for some cuts. Owen's got a few thousand left on him, if we feel like playing."

"Okay," Victor said, but Semyon could tell from his voice that he was less than thrilled about Semyon and Owen wandering off alone.

Semyon shrugged, rising back to his feet. Victor was the boss when it came to blackjack, but that didn't mean he was going to run their lives. Semyon hurried out of the stall, making sure to move fast enough so that he and Victor didn't cross paths on the way out. It wouldn't look good for them to be seen coming and going from the bathroom together, if anyone was watching. And he didn't need to get a look at Victor's face to see the displeasure that went along with his cautionary tone. Semyon didn't need another father figure in his life. One was more than enough.

He hit the bathroom door with his shoulder and reentered the casino floor. The casino was crowded, and it took him a good ten minutes to work his way past the table games to the huge bank of slots that filled the vast entranceway of the dramatically decorated casino. Like everything else in the Trump empire, the Trump Plaza was violently overdone, a study in decorating hubris rivaled only by the ego of the man behind the casino himself: crystal chandeliers, velvet carpeting, faux-antique furniture along the paneled wooden walls, and way too much gold trim. Semyon couldn't begin to imagine how much money had been spent on the casino floor. It made him feel good inside that he wasn't contributing a penny to help pay for all the garishness.

He reached the front entrance and passed through a huge revolving glass door. He felt the air outside was muggy and warm as he took the steps that led down toward the street. He had to push through a group of men in business suits at the bottom of the flight of stairs, then somehow found himself in the midst of a gaggle of Jersey girls, frizzy hair sprayed high upon their heads, cutoff jeans and matching jean jackets, accosted by a cloud of perfume so strong it made him gag, and somehow he got turned around, finally finding himself on the curb, searching for the way that would lead him toward the Boardwalk. He couldn't see any street signs and wasn't sure which direction he was supposed to go. He'd told Owen to meet him right outside the casino, but after he'd scanned the area, it was clear Owen had either gone on ahead or was waiting for him somewhere else.

He was about to cut back toward the entrance to get his bearings when he felt a hand close tight around his right wrist. He looked up and found himself staring at a strangely familiar face. Then he placed the slicked-back black hair, colorless lips, and shitty brown suit— and his stomach turned over. It was the Latino guy, one of the two men who had been watching him playing in the casino.

"Hey, man, nice run you had going on there at the blackjack table."

Semyon tried to yank his arm free, but the man's fingers were closed around his wrist like a vise. Before Semyon could say anything, he heard a screech of tires and watched as a car pulled to a stop right in front of him, almost up on the curb. The car was an American make, light blue and at least ten years old, with tinted windows and dents running up and down its body. It sat low on its rear tires, or maybe it was the front ones that were jacked up, gangster style. The driver's window was open, and the other brown-suited man from the casino was leaning out, one arm hanging down against the door. Semyon's eyes drifted to the man's hand because it was holding something dark and metallic.

Semyon didn't know much about guns but had seen enough movies to recognize a .35-caliber automatic, especially one that was pointed at him from just a few feet away.

"Get in the car, motherfucker."

Semyon's eyes widened as he looked from the gun to the man holding his wrist. He couldn't believe this was happening. There were people all over the place. The Jersey girls had gone into the casino already, and the men in business suits had turned the corner, but there were people right across the street, and more passing through the revolving door at the top of the stairs. Still, nobody was close enough to see the gun, or the grin on the Latino's colorless lips.

"I ain't gonna tell you again."

Semyon's feet felt like they were glued to the sidewalk. The Latino yanked open the back door of the car, then gave his wrist a violent tug. Semyon wanted to resist, but he couldn't break his eyes away from the barrel of that gun. He'd never had one pointed at him before. He wanted to scream, to run, to try anything to avoid getting into that car—but that dark gun barrel was now doing the thinking for him. And it told him he didn't really have a choice.

He took a step forward and followed the Latino into the backseat of the car.

The fist came at him low and hard, catching him at the base of his lower jaw, sending his head jerking back into the passenger-side window with enough force to make his ears ring. He immediately tasted blood, and his hands went up, trying to protect his face from the next blow. But he was too slow; the fist hit him again, this time on his right cheek, sending his head twisting to the side. He cowered back against the car door, trying to think past the shards of pain. Fuck, this was bad.

He opened his eyes and caught a glimpse outside the window; they were moving fast, maybe fifty miles per hour. Streetlights flashed by, but at irregular intervals; they were no longer near the Boardwalk, they were heading deeper into the inner city—the jungle that Victor had so helpfully described. Semyon could hear the roar of the car's engine mingling with the squeal of the tires as they took corners too fast. He had no idea where they were going, but it didn't really matter. He wasn't sure he was going to survive the ride.

The man with the narrow face was still driving, but he'd handed the .35 to the Latino, who was holding it loosely in his left hand while he beat the crap out of Semyon with his right. This had been going on since they'd pulled away from the curb, maybe ten minutes ago.

"Christ," Semyon cursed, still covering his face with his hands. "What the fuck do you want from me?"

He felt blood dripping down his chin and was pretty sure his bottom lip was swelling up from the beating. There was a knot growing on the back of his head where he'd hit the window, and one of his upper teeth felt slightly loose against his tongue. Otherwise, he was still in fairly good shape, but wouldn't be if this kept up much longer. Still, it was better than getting shot. If he tried to fight back, he knew he'd end up dead. So he had no choice but to figure out what the hell the two fuckers wanted and give it to them.

"What do you think we want?" the Latino shouted back. He made a fist again—but held back, instead pointing with the .35. "Now give us the fucking money."

Semyon lowered his hands a few inches, looking from the gun to the man's face. The man's lower lip was twitching, and his eyes were red around the pupils. Definitely on drugs, maybe even a junkie. Shit, Semyon finally understood what was going on. These two didn't work for the casino, and they weren't private eyes. They had been casing him at the casino and had seen him winning big. They had waited outside for him—to rob him. Shit. If they had worked for the casino, at worst, they would have just tried to scare the shit out of him. These guys—they might actually kill him.

Semyon held out his hands, palms out, then slowly reached into his pants pockets and pulled out the three hundred dollars that he hadn't handed over to Victor.

"This is all I've got. I swear."

The Latino snatched the bills from his hands, then looked at them and tossed them to the floor.

"You lying motherfucker! We were watching you the whole time you were playing. You must have won a hundred grand at that table."

Semyon felt like he was going to throw up.

"I don't have it on me, man. You think I'd walk outside of the casino carrying that kind of cash?"

The driver swung his head around, spitting bile. "He's fucking lying. Search the motherfucker!"

Semyon held up his hands. "Go ahead! Search me!"

The Latino reached forward with his free hand and began patting Semyon down. Semyon tried to ignore the hard fingers grabbing at his chest, down the sides of his shirt, even into his pants. When the man was finished, he leaned back, and there was real fury in his dark eyes.

"You commie fuck. Now you're dead. You're fucking dead."

Semyon hadn't been called a Communist since high school. Back then, he'd been called it almost every day. But in the early eighties, it had made a little more sense. For some reason, the word jabbed at him, made him angry. He was still terrified, but the anger pushed just enough of the fear aside to give his mind some clarity. He had been in street fights before, had known people like these two junkies from the streets where he'd grown up. He knew that the next few seconds would determine if he lived or died.

Somehow, he calmly lowered his hands and looked the Latino right in his eyes.

"So you're going to kill me—for nothing? You're going to get yourself a homicide rap, get yourself the fucking gas chamber for whacking some Russian dude for three hundred lousy bucks? You want to get a death sentence for *this*?"

The Latino stared at him. The driver was glancing in the rearview mirror, and Semyon could see the panic in his eyes. This hadn't gone down the way they had planned it—if they'd planned it at all. They'd hoped for an easy score. Armed robbery was one thing—but murder, another. Semyon saw his advantage and summoned a moment's more courage. He licked the blood off his lower lip and crossed his arms against his chest.

"Look, right now I don't know who you guys are, and you guys certainly don't know who I am. Because if you did, we wouldn't be having this conversation at all. A Russian, betting the kind of money I was betting. You think I made that money selling vodka? You think my associates aren't going to come looking for me? You want to get yourself in deep—for *this*?"

The Latino glanced at the rearview mirror, meeting his partner's eyes. Then he exhaled.

"Fuck."

His decision was made. Without another word, he reached past Semyon with his free hand, viciously grabbed the door handle behind him, and pushed it open. A blast of muggy air rushed into the car, and before Semyon could open his mouth to scream, he was rolling backward, right out onto the street.

His shoulders hit pavement and then he was tumbling, heels over head. He made three full revolutions before coming to a stop, lying flat on his back, head against the curb, staring up at the dark sky.

He blinked, mentally checking his limbs. Nothing seemed to be missing, nothing seemed to be broken. Obviously, the driver had hit the brakes right before the Latino had pushed him out of the car, otherwise he'd be dead, or at the very least missing some of his important parts. He didn't know where the fuck he was, and his face hurt like hell, but he was alive. He'd survived having a gun pointed at him, for the first time in his life.

He prayed to God it was the last.

CHAPTER 16

The Trump Plaza
Atlantic City, New Jersey

The water was ice-cold, but burned like fire as Semyon dunked his head into the oversize, gilded sink. Even though his eyes were tightly closed, the soap stung at him, but the sudden pain was good, making him forget about the throbbing in his skull and the raw ache of his face.

He kept his head submerged as long as possible, until the last seconds of air had been used up by his lungs and he had no choice but to face the world. Then he pulled his head out of the sink and let the cold water run down his neck and soak his blood-spattered white T-shirt.

"Man, you look like shit."

He glanced at Victor in the backlit, bathroom mirror. Victor was seated on the closed, thronelike toilet seat, legs crossed at the knees, his arms tight against his chest. He'd taken his suit jacket off and rolled up his sleeves when Semyon had first entered the four-room suite—to avoid getting blood on his clothes as he helped Semyon clean himself up—and he looked surprisingly naked in his off-white, tailored shirt and thin gray tie. He also looked surprisingly small against the backdrop of one of the biggest bathrooms Semyon had ever seen. From the massive, boatlike bathtub in one corner to the dual glass-walled showers in the other, the place looked more like a marble-tiled sanitorium than a bathroom. There were mirrors everywhere—which, at the moment, was not a good thing, considering Semyon's appearance. The forty-minute walk back to the casino hadn't helped matters; he wasn't just beat-up, he was also exhausted, and his shoulders slumped forward with the very effort of keeping his head atop his neck. But it didn't do any good to dwell on his misery. And he didn't really want Victor's sympathy—because he didn't really think it was all that sincere. Although Victor's features showed concern, his eyes communicated something else. Semyon had seen that same look when Victor was running the second technique, predicting the location of an ace by the sequence of cards that came before it. It was as if he was implying, through those eyes, that he'd predicted something like this might happen. Which, of course, was ridiculous.

"Could have been a lot worse," Semyon mumbled, shaking the water out of his ears. He surveyed the damage in the mirror. His lower lip was pretty swollen, and there was a bright red mark on his cheek where the man had hit him. Other than that, with most of the blood washed away, he didn't look nearly as bad as he felt. His head still ached from the window, and his back was covered in bruises from where he'd hit the asphalt. His elbows and knees had fairly deep scratches, and he'd probably need to see a dentist about the loose tooth in his upper jaw when he got back to Boston. But otherwise, there didn't seem to be any real damage. He'd been damn lucky.

"Yeah, you could have still had our money on you. Then we'd be out more than a hundred grand."

Semyon stared at Victor in the mirror. Victor smiled, holding up his hands.

"I'm kidding, man. It really sucks what happened to you. I told you this city was dangerous. None of us should wander off alone."

"I wasn't wandering off alone," Semyon shot back. He felt himself getting angry. He definitely didn't want a lecture at the moment. He'd never talked back to Victor before, never really had a reason to. But he'd just had the shit kicked out of him, and the last thing he needed was more fatherly advice.

"That's right," Victor said, rubbing his hands against the shiny material of his pants. "Owen was supposed to be with you. What the hell happened to him?"

Semyon shrugged, gingerly touching his swollen lip. "We must have missed each other. I think I took a wrong turn outside of the casino. Anyway, it happened so fast, there wouldn't have been much he could have done to stop it."

Victor didn't respond. He went silent for a few seconds and seemed to be pondering something. Then he sighed and reached into his pants pocket. He pulled out a sheet of folded paper and held it out toward Semyon.

"I was going to show you this before we left for Atlantic City, but decided to wait. I didn't think it really mattered—I still don't. But I want you to see it."

Semyon turned away from the mirror, wondering what the hell Victor was going on about. His shoulder hurt as he leaned forward to retrieve the folded sheet of paper. He needed to lie down, get some serious rest. There would be no more gambling for him this weekend, that was for sure.

He unfolded the sheet of paper. The page was covered in dark type and looked to be a photocopy of some sort of official document. On the top of the page, there was a heading that identified the paper as being the property of the police department of Hartford, Connecticut. Beneath that, a line of type that Semyon couldn't quite make out because the photocopier had blurred some of the important words. Below that, however, was a name he could easily read, one that was extremely familiar. *Owen Keller.*

"What am I looking at?" Semyon asked.

"It's a police record. Owen's police record. Jake used some of his law-school connections to do a background check, and found this little gem. Seems that Owen was arrested three years ago for possession of narcotics. Did three months in the Hartford Penitentiary. He must have written one hell of an admissions essay to get into MIT after that."

Semyon stared at the paper in his hand, then shrugged, refolded it, and held it out for Victor.

"So?"

Victor took the paper, shoving it into his pocket.

"So nothing. I just wanted you to be aware."

He hit the last word hard, and Semyon found himself getting annoyed. He gestured toward his bashed-up face.

"You think this was somehow Owen's fault?" he asked. "Because he wasn't standing right outside the door, waiting to hold my hand? I got turned around and ended up getting mugged. Owen had nothing to do with this."

"I'm not saying he did," Victor quickly responded. "I'm just saying you should be aware, that's all."

Semyon turned back to the mirror. He didn't want to continue this conversation. He didn't care that Owen had once been arrested for drugs. Sure, sometimes it seemed like Owen might be on something—but he was a good blackjack player, and also a good friend. If he'd had a real problem, it would have shown up by now. And he certainly didn't have a malicious bone in his body. Maybe he'd fucked up in the past—but he was a good kid. Semyon trusted him. Which was more than he could say for Victor, at the moment.

A thought hit him, and his eyes narrowed. He looked at Victor, sitting on his toilet throne.

"So did you and Jake run a background check on me, too?"

Before Victor could answer, there were sounds from another part of the suite—a door opening, multiple voices, laughter, drunken and happy. It was the rest of the team, returning from a very good night of gambling. It sounded like they were all there, Allie, Owen, Jake, Misty, and Carla. Moving through the suite, room by room. From the massive living room, with the picture window that looked out onto the ocean, to the galley kitchen, with a fully stocked refrigerator, toward the bathroom.

Semyon had no choice but to let the matter drop as Allie reached the bathroom first. She had Owen's jacket on over her T-shirt and was laughing at something someone had just said. Then she saw Semyon and her face froze, her mouth wide open.

"My God, what the hell happened to you?"

Owen was right behind her. He stopped dead in the doorway.

"Jesus Christ, man. When you didn't show up at the Boardwalk, I assumed you'd changed your mind and decided to stay for a few more rounds of blackjack. It looks like you went a few rounds with Mike Tyson instead."

Carla and Misty pushed by Owen, then rushed straight to Semyon, going to work with wet towels and more soapy water. Semyon tried to stop them, but it was no use. Allie joined in, and pretty soon he was wincing as they finished the cleaning job he'd started. They'd pretty much done as much as they could do by the time Jake had lumbered into the bathroom, making the scene complete. Quite a bathroom, that could fit an entire cabal of MIT card manipulators. Donald Trump would have been proud.

"I guess we should just look at this as an important lesson," Jake commented after Semyon told them what happened. "We've got to be more careful. We're dealing with a lot of money. We've got to be extra vigilant, especially when we're traveling around to places like this."

Victor nodded. "Travel is something we'll definitely have to look into. If we're going to hit AC on a more regular basis, I've got some ideas of how we can make things more . . . refined."

Semyon raised his eyebrows. The motion hurt, and he let Allie ask the question for him.

"What sort of ideas?" she asked. But she didn't seem all that interested in Victor's answer. She was looking at Semyon's bruised face, real concern in her eyes. He had the feeling she wanted to touch him, give him some physical assurance that he was really all right, but she wouldn't do it, not in front of the entire team. Still, the thought took the edge off his pain, made it almost bearable.

"You'll see," Victor responded. Then he shifted gears, rising from the toilet seat.

"But first, we're going to take the incredible profits that we've made so far, and go on another, even more profitable excursion. Really take things up a notch, so to speak."

Owen hiked himself up onto the counter by the sink. He glanced at Semyon, winced along with him, then turned toward Victor.

"We heading back to Vegas, boss? Parade Semyon around, show

the boys how to really work someone over, AC style? None of that handcuff-to-a-chair, pussy shit?"

He was making a joke, but Victor didn't seem amused. Victor shook his head.

"We're not going back to Vegas yet. And we're going to take a break from AC, at least for a little while. I'm talking about someplace much more restful. We've earned ourselves a vacation, don't you think? Semyon, most of all."

"What sort of vacation?" Misty asked.

Victor grinned.

"Paradise, baby. I'm talking about paradise."

CHAPTER 17

A Casino
Aruba

The first thing Semyon noticed about paradise was that it was unbearably hot. Not the moist, muggy hot you expected from a Caribbean island on the westernmost edge of the Dutch Antilles, but an arid, desert, dusty hot that smelled and tasted like Las Vegas.

The second thing Semyon noticed about paradise was that it was windy. Not a comfortable wind that ruffled gently through his resort-casual outfit—short-sleeved, flowery shirt, comfortable white slacks, open-toed sandals—but a mean, angry, Venezuelan wind that whipped against his skin, aggravating the fading, week-old bruises on his lower lip and cheek.

The third and most surprising thing Semyon noticed about paradise was that it was full of goddamn cacti.

Christ, from what Semyon had seen during the short and bumpy ride from the minuscule Aruba International Airport, the tiny island was about 2 percent beach and 98 percent cacti. Not the little, spiky numbers you saw in Nevada; these were big-ass monsters, looming fifty feet high above the craggy, rocky sand. They seemed to grow in fields so dense, they looked like they had been purposefully planted by some maniacal and sadistic farmer.

"Doesn't quite look like the brochures," Owen shouted from the seat directly behind Semyon, his voice barely audible over the draft that was tearing through Victor's rented Jeep Cherokee. "Where are the white sand beaches? The palm trees? The fucking cabana girls in string bikinis?"

Semyon grinned at him in the rearview mirror. Owen was clinging to the Jeep's roll bar with both hands as Victor navigated the four-wheel-drive vehicle over the winding island road. Allie, in a short skirt and a painfully skimpy bikini top, was jammed into the back middle seat next to Owen and was holding on to Jake, his massive bulk crammed in next to her, for dear life. Semyon was lucky to have the front passenger seat all to himself, since Carla and Misty had remained back in Boston. Something to do with an applied-math seminar they were making up so that they'd be able to graduate during the fall.

Semyon shifted his gaze from the mirror to Victor, who was consulting a half-folded map with one hand while fighting with the oversized steering wheel with the other. Victor's eyes weren't visible behind a pair of wraparound mirrored sunglasses, but his small, feline body seemed relaxed beneath his cream-colored leisure suit. Semyon wasn't sure where he had found the suit, which seemed more suited for a Miami drug lord than an MIT cardplayer, but maybe that was on purpose. Miami drug lord might make a good casino character.

"Oh, there are beaches," Victor finally said as he shoved the map back onto his lap and grabbed the wheel with both hands. "They're on the western side of the island. But this side, the east, is pretty much what you see. Desolate, windswept, covered in strangely shaped rocks and cacti. Now if that's not paradise, I don't know what is."

Semyon grabbed the dashboard as they took a hairpin turn. There was a cliff to his right, a high, jagged stone wall that led all the way down to the ocean. The water was blue, almost transparent, and the view out toward the horizon was pretty good. It was bright out—midmorning, probably closing in on noon—and in the distance, he could barely make out the mountainous Venezuelan coast. He had glanced at a guidebook in the airport during their layover in Miami and knew that Aruba was a bare twelve and a half miles from the South American country. Twenty miles long, six miles across, best known for its eleven or so casinos and its handful of oil refineries, the island was sometimes referred to as the Vegas of the Caribbean. Semyon only hoped the island was a bit more friendly than the Vegas of the East.

They took another tight curve and began to descend a small hill. A cluster of low-roofed, pastel-colored buildings came into view, huddled together around a fairly cluttered port. Picturesque, if a little down-market. As they moved closer to what passed for the island's capital city, Semyon noticed a number of pedestrians strolling along the dirt road. Brown-skinned, thin-limbed, most in cheap, poorly fitting clothes—low-hanging shorts, wife-beater tees, native island wear. Like most islands in the Caribbean, Aruba was a place that catered to wealthy, foreign tourists—while the majority of the island's inhabitants were actually living well below the poverty level. Rich white men and women dined on lobster and margaritas, while the natives scraped by, hoping for a little bit of the old trickle-down. It didn't help that the island was bolstered, primarily, by a casino economy. Aside from Vegas, the one glaring aberration to the form, "casino economies" usually worked out pretty good for the casinos, and not so good for the local economy. The trickle down theory didn't really apply to slot machines and blackjack felts.

Semyon smiled inwardly as Victor gunned the Jeep's engine, speeding their way toward the pastel town. At the very least, the MIT crew would be striking a little blow for the common man.

If the drive from the airport had been lacking in postcard moments, the short boat ride to the hotel resort more than made up the difference. To Semyon's surprise, the little motorized ferry had actually picked them up right in the middle of the lobby of the full-service resort. A small canal had been cut right through the center of the hotel's main building, leading out to the lapping waters of the capital city's small port. The boat ride to the private island was about ten minutes long, a cute gimmick that made Semyon almost forget about the cacti and Venezuelan wind as he surveyed the resort where they'd be spending the rest of the weekend.

The place was pretty ritzy, located on a private stretch of beach on a private island and completely full-service. Three pools, fifteen restaurants, a movie theater—and two megacasinos, one right up on the beach, the other opposite the port. But the casino they were interested in was off-island and not part of the resort. Victor had assured them that this casino could handle some pretty heavy action. It was a lavishly decorated glass complex that was bristling with marble, brass, and gold leaf, as well as enough dripping-crystal chandeliers to make Donald Trump jealous.

They had been comped two adjoining suites in the main tower of the resort; although Victor didn't have a dedicated host at the hotel's casino, one of his aliases—a commodities trader from Manhattan named Danny Pesto—had a good enough relationship with a sister casino on another island in the Caribbean to get them the VIP treatment. The suites were pretty impressive, with oversize balconies overlooking the ocean, and wide, wraparound picture windows. Both had circular living rooms with pastel carpeting, rattan furnishings, and even hammocks slung from brass stands in alcoves by the windows. They all had their own bedrooms, and four bathrooms to split between the five of them. Still, they didn't expect to be spending much time at the hotel.

After a quick lunch of room service—fresh seafood served on one of the balconies by a uniformed waiter with a pasted-on, super-white smile and the manners of an English butler—they prepared themselves for their first assault on the nearby casino.

Semyon had already settled on his character for the weekend. He would play a rich, prep-school-bred, trust-fund brat named Parker Buckingham—the kind of kid he'd often seen wandering around the Harvard campus and had once imagined Allie dating—before he'd gotten to know her better. To complete the transformation, he'd parted his hair severely, right down the middle, and tied a white cardigan around his shoulders. He'd also brought a prop with him—a graphite tennis racket that he'd borrowed from one of Jake's friends at law school. Although he felt a little strange carrying the tennis racket with him into a casino, Victor had assured him that it was exactly the sort of thing a rich kid like Parker Buckingham would do.

Owen was going to take the playacting one step further; instead of a cardigan, he was wearing a tight pink Izod shirt, complete with alligator on the lapel—and had traded his ripped jeans for a pair of white Bermuda shorts. And instead of a tennis racket, he was clutching a brightly colored, oversize plastic water gun, a device called a Super-Soaker that could spray water in a twenty-foot arc. "Chad Helms" was more than just a rich kid like his buddy Parker Buckingham; he was a bit of an asshole and liked to make a scene wherever he was. Owen was going to be perfect for the role.

Allie didn't have to stretch far to inhabit her own character: Lilly Bowles, an ice princess and heir to a textile fortune, blowing her daddy's money while slumming it in the Caribbean instead of off on her family's yacht in the French Riviera. Jake, in a dark suit and sunglasses, was posing as her bodyguard, and Victor, in his cream suit, with his Danny Pesto ID, was her wealthy New York boyfriend. If they were pressed on their overall backstory, they'd all met at a party during the Montreal Grand Prix. But Victor doubted anyone would be questioning them once they started throwing down the hundred-dollar bills. The Caribbean wasn't like Vegas; though a few of the casinos could handle the action, they weren't the bloated business conglomerates of the Vegas Strip. Most of these casinos were privately owned and hungry for the kind of money that Victor's team was carrying.

"Okay," Victor said as he surveyed the group, lined up on the balcony, blue water behind them, the glass casino glowing on the horizon. "Let's show this little island what a bit of math, in the right hands, can do to balance out a few hundred years of economic oppression, shall we?"

Semyon grinned, and barely felt the pinch of his still bruised lower lip. *Robin Hood had nothing on them.*

○ ◎

"Pay the table!"

Owen was up on his feet, the Super-Soaker high above his

head, a thin arc of water spraying out over the crowd of tourists that had gathered behind their table. The crowd couldn't decide whether to applaud or duck for cover as they watched the dealer begin counting out forty thousand dollars in brightly colored chips. Semyon was laughing so hard at Owen's act he was nearly falling off his stool. Allie was also on her feet, clapping her hands as Victor hugged her around the waist. Only Jake kept his calm, seated quietly at his first-base seat, arms crossed against his oversize chest.

After only two hours in the bustling casino, they were way up—maybe seventy or eighty thousand in total, half of which they'd won in this single application of the third technique. In that time, they'd switched tables twice; the first dealer, an old British gentleman with huge hands, had given them a few opportunities for the second technique, but the ace sequencing hadn't amounted to any significant profits. The second dealer, a young woman with long fingers and a cumbersome engagement ring, had offered up two ace cuts, for a few thousand in winnings. But it was this third dealer—a thin black man with a heavy island accent and a charming smile—who had really broken things open for them. After each shuffle, he didn't even seem to make the slightest effort at covering the back card as he rolled over the shoe—and why should he? Obviously, the casino had no idea that the last card could be so significant.

Semyon reminded himself that they still had to be careful, that they couldn't get too greedy. He, Victor, and Jake had quickly discussed things in the bathroom ten minutes earlier and had decided that they'd hit the casino for only another hour or so before taking the rest of the night off. They didn't want to win too much, after all, and raise the casino's suspicion level. So far, the two uniformed pit bosses who were watching the display from an elevated perch above the blackjack pit didn't seem to be overly concerned about the group of American rich kids making a scene at the corner table. In fact, they probably thought the excitement Semyon's team was generating was good for business, evidenced by the huge crowd of onlookers that had gathered to watch their megabets. Moreover, the pit bosses hadn't even admonished Owen for the Super-Soaker, not wanting to interrupt the big-money play.

Semyon turned his attention back to the table. The dealer had finished paying the table and was getting ready to deal out the next hand. Semyon quickly lowered his bet back down to the minimum—three hundred dollars for each of his two hands—expecting the others to do the same. Then he glanced over toward Jake—and saw that the big man had his hands crossed against his chest.

Semyon blinked. Midplay, hands across the chest meant put the big bet in again; in other words, Jake had either tagged an ace that was predicted to come out in this round or another ten card to manipulate toward the dealer. Either one was quite possible, considering how sloppy the dealer was being.

Semyon spun his tennis racket in his hands, then cleared his throat.

"What the hell. Let's keep things exciting."

He reached forward and pushed the big bet back into the betting circle. He watched as the rest of his team followed suit. Then he glanced at Jake again, who had spread his hands out against the felt— the signal that he'd pegged another ten card, the third technique. Semyon wasn't sure how or when he'd done it, but he'd obviously watched a second ten through the dealer's shuffle. Semyon wasn't sure how many cards they would have to take to steer the ten into the bust position, but he was certain that Jake and Victor would be able to control the situation. In the end, that was what it was all about, controlling the table. And at the moment, they had complete control of the table.

"Here we go again," Owen said, aiming his Super-Soaker at the crowd.

The cards came out. After the first part of the deal, Semyon was sitting on a seventeen and a fourteen. Owen had a sixteen, Allie a twenty, Victor a fifteen, and Jake a nineteen. The dealer was showing a four, which meant that if indeed Jake had pegged a ten, there was a very good chance it would be the bust they were looking for. It was just a matter of taking away the right amount of cards to steer the winning card to the right place.

Jake waved his big hands, indicating that he didn't want a card, and the dealer shifted toward Victor. Before Victor could do anything, Jake tapped his fingers on the table.

"Fifteen, eh, Mr. Pesto? Against a four? I think the book says you should hit that one."

His statement was blatantly incorrect, but Victor simply smiled and pointed at his card.

"The man knows his game," he said. There were hushed voices from the crowd behind—obviously some basic-strategy players were in the house and were commenting on how stupid the rich people at the table were being—but Victor ignored them. "Give me a card, boss. And make it a small one."

The dealer turned over a nine, busting Victor's hand. Victor cursed, sighing. Then the deal went to Owen and his sixteen. Again, Jake cleared his throat.

"Gotta go by the book, man. Gotta hit that sixteen as well."

Semyon could see the pitying look on the dealer's face. He knew that Jake was giving bad advice, but of course he wasn't going to say anything. Not with fifty thousand dollars still on the table. Owen shrugged, pointing at his card with the Super-Soaker.

"Fucking gimme something I can use, man."

The dealer turned over a two, for a hard eighteen. Not a bad pull, after all.

Now it was Semyon's turn. He looked at his seventeen and his fourteen, then suddenly turned toward Jake.

"Well, Mr. Know-It-All. What does your fancy book say I'm supposed to do with these hands?"

Jake smiled at him.

"Heck, what do I know. Why don't you just stay on both of 'em and leave it up to God."

Semyon shrugged. It sounded like good advice to him. He waved his hands.

The dealer turned over his down card, a nine, for a thirteen. Then he reached into the shoe and withdrew a jack of hearts. The bust card.

And Owen was back on his feet, the water gun in the air.

"Pay the table!!!"

The crowd roared as the dealer pushed fifty thousand dollars in chips into the betting circles. Now they were up one hundred and twenty thousand dollars. Semyon was smiling so hard it hurt. He was about to pull back his bet, to start again—when he felt Owen kick him under the table. He looked up and saw that both Allie and Victor were running their hands through their hair, simultaneously. It was the first signal that Victor had taught them. And it had nothing to do with the game.

It had to do with the four security guards who were suddenly pushing through the crowd, heading straight for their table.

Semyon didn't even pause. He scooped up all his chips and jammed them into his pockets. Out of the corner of his eye he saw the rest of his team doing the exact same thing. The dealer, caught by surprise, started to say something about coloring up the chips—but Semyon was already stepping back from the table. Shit, they had to get out of there. Fast.

He saw Owen breaking away first, losing himself rapidly into the crowd. Allie was a step behind him, moving almost as fast. Semyon tried to follow—but before he'd made it two steps, he felt rough hands grab him by the shoulders and yank him back. A second later he was pinned up against the blackjack felt. Victor and Jake were next to him, both held by burly security guards. The guards were all dark-skinned, probably natives of the island, and all wearing matching uniforms. The biggest of the three had a big smile on his face. His accent was rhythmic and heavy.

"So where you gonna run to, man? It's an island."

Semyon had to admit, the man had a pretty good point.

As back rooms went, this casino's was actually pretty comfortable. There was cheery yellow wallpaper on the walls, a good air-conditioning system, and even a small window that looked out on the ocean. The chairs on which Semyon, Victor, and Jake were sitting were cushioned and modern, and even though the three guards had followed them into the room, closing the heavy wooden door behind them, the place didn't feel all that cramped. Still, Semyon could not help the fear that rose inside his stomach as he looked at the three guards, who were now chatting away in one corner of the room. They hadn't done anything menacing, yet, but only a week ago Semyon had been beaten near senseless, and he really didn't want to repeat the experience.

The good news—beside the fact that Owen and Allie had gotten away, for the moment—was that Semyon still had his chips in his pockets; the guards hadn't asked any of them for the money, which meant that the casino wasn't sure, yet, what to do with them—or if they'd really done anything wrong. Obviously, somebody had gotten suspicious when they'd won the second big hand in a row, but maybe that's all it was, suspicion. If they played their cards right, maybe there was still a way out.

Finally, the three guards finished their conversation. The big one, with the bright smile, turned toward them.

"You guys are very lucky, eh?"

Victor leaned back in his chair, hands behind his head.

"Why, because we won two hands? Come on, man, this is bullshit."

"Maybe too lucky?" the guard continued. "Maybe, you got some sort of computer on you, make you so lucky?"

Semyon almost laughed out loud. So that was it. The security that had obviously been watching them through the cameras knew they weren't card counters. But they also knew they were winning, which seemed impossible. So they'd come up with the brilliant idea that one of them must be wearing a computer. It was ridiculous. What sort of computer could predict a ten card?

Jake was less controlled than Semyon and guffawed right out loud.

"What, you want to do a search? You want to see if we're wearing some electronics?"

He stood up out of his chair. The three security guards stared at him as, suddenly, he started to undo his pants.

"Is that what you want? Do a strip search?"

With a flourish, he yanked his belt open and let gravity do its work. The three guards gasped, eyes wide. Semyon blinked, equally in disbelief. Not only were Jake's pants now down around his ankles—but he wasn't wearing any underwear. His naked body glowed white in the light from the window.

"Any of you see a computer?"

The guards had backed as far away as they could. Semyon looked at Victor, who was also staring wide-eyed at Jake. Before any of them could say anything, the wooden door opened and a man entered the room. He was tall, elegantly dressed in an oxford shirt and pressed tan slacks, with wavy brown hair and smooth, tan features.

"I think you've made your point," he said simply. "You can put your pants back on."

Victor waited until Jake had sheepishly pulled his pants back over his waist before speaking.

"Are you the casino manager?" he asked. It was a pretty good guess, though the man looked too polished to be a manager. He was almost model handsome, probably in his midfifties, and seemed like he'd just stepped off the deck of an expensive yacht—or out of the pages of a J.Crew catalog.

"My name is Michael Sterling. I'm a shareholder in this casino, actually, meaning I pretty much own the place."

He smiled, and Semyon saw that his teeth were capped. Victor was smiling right back at him.

"Are you here to break our knuckles?" he asked. Semyon glanced at him, wondering if he was pushing it. God only knew what sort of people owned Caribbean casinos. But it seemed as though Semyon's concerns were unnecessary.

"Actually," Sterling responded, "I'd like the three of you to join me for a round of golf. So we can discuss this like gentlemen."

Victor glanced back at Semyon, then looked at Jake. They didn't need to discuss the offer because it certainly beat any alternative. Victor answered for all of them.

"As long as my friend here can keep his pants on, you've got yourself a foursome."

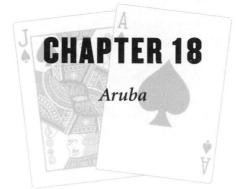

CHAPTER 18

Aruba

emyon wasn't sure who looked the most out of place on the manicured green. Victor, in his cream-colored suit, holding a golf club gingerly in the fingers of his right hand like it was some sort of alien artifact. Jake, sweating profusely even though his black suit jacket was off and balled up under his arm and his shirt was open three buttons down the front. Or Semyon himself, bending over the little white, pockmarked ball, his own club gripped in both hands like a hockey stick over a puck, imagining what it would be like to actually be able to hit the damn thing.

The only one who looked like he belonged on the green was Sterling, but he was a member of the golf club, so that wasn't really a fair contest. He was watching Semyon's efforts with some amusement from a spot at the edge of the grass. Behind him, a gentle hill led down a sandy expanse to what looked to be an elegantly appointed clubhouse. Next to the clubhouse was a parking lot, where Sterling's limo was still waiting—parked next to a pair of island police cars. The police cars had escorted them to the golf course from the casino, and Semyon wasn't sure whether the police presence made him feel more comfortable or less. After all, whom did these island police work for, if not the man who brought in a good percentage of the country's gross national product?

Semyon took a deep breath and decided he was better off concentrating on the golf ball than on the situation they were in. He gripped the club even harder, leaned back, and gave it a hefty swing. There was a whiff of air— and he nearly toppled over as the club went past his shoulder. The ball remained right where he had put it, unscathed.

"I guess we can't be good at everything," Sterling said as Semyon dejectedly put the club back in Sterling's bag. "You guys should definitely stick to cards."

Victor poked at the grass with his club. He'd had about as much luck at the game as Semyon. Only Jake had managed to hit one of the balls, and it had gone off at such an angle that it had nearly sailed over the parking lot.

"What makes you think we're any good at cards?" Victor asked. "I mean, we get lucky like anyone else—"

"Don't insult me," Sterling said, though he still sounded more amused than angry. "I was watching your play for an hour, maybe longer. I know you're not card counters—I've known how to count since the eighties, actually, and can spot a counter a mile away—but I also know you're not regular players. And you're definitely not who you say you are."

Victor shrugged, and the man turned his attention toward Semyon.

"I got curious, so I snapped some stills of your faces with our cameras and faxed them along to an associate of ours out in Las Vegas. A man named Jack Galen. I think you're familiar with him, Dr. Nussbaum?"

Semyon's cheeks twitched. He glanced at Victor, then back at Sterling. Shit. If Sterling had talked to Galen, then he knew that they weren't a bunch of rich kids on holiday.

"It's funny," Sterling continued. "Every year we get a few guys coming down here with systems they've worked out to beat the house. Usually we let them go about their business. They win for a little while, then their luck changes, and they inevitably lose their stake. Because none of the systems work. Other than the card counters, and the cheaters, everyone else eventually loses."

He stepped forward and retrieved the golf club that Semyon had used out of the leather bag.

"But I could tell right from the start that you guys are different.

Every time you bet big, you won. It's almost as if you knew when the good cards were coming."

Semyon's throat felt dry, and it wasn't just the arid air.

"Give us a little more time," he chirped, surprised by his own voice. "I'm sure we'll eventually lose, just like everybody else."

Sterling glanced at him. Then he crossed the green to the ball and moved into a perfect golf stance. He swung the club hard, connecting perfectly, and the ball sailed out across the course. Semyon lost sight of it as it hit its apex, but he was sure it had landed somewhere near the hole.

"I don't think so," Sterling said. "Mr. Galen assures me that you're up to something. He doesn't know what, but one day he's going to find out."

Jake rubbed sweat off the back of his neck with his palm. "So what do you want from us, Mr. Sterling?"

Sterling paused, looking at each of them. Then, in one swift motion, he reached into his pocket and pulled out a gun. Semyon's eyes went wide. For the second time in two weeks, he was staring down the barrel of an automatic.

"I want my money back. Every penny of it."

Semyon gasped. He couldn't believe this was happening. A second ago they were playing a friendly game of golf. Now the man was robbing them. Well, not robbing them, exactly. Demanding back the money that they had won. It was absurd. The man was most definitely a multimillionaire. He owned a casino and maybe even the golf course. But Semyon realized that it probably wasn't just about the money. It was the principle of the thing. They had beaten him, in his hometown. In his house.

What were they going to do? They were on an island twenty miles from Venezuela. Sterling was a casino owner with obvious ties to the island's government—and he was holding a gun.

Semyon looked at Victor. Then at Jake. Without a word, all three of them began to empty out their pockets. Sterling watched as every last chip fell to the manicured grass. Then he gestured with the gun, toward the police cars in the parking lot.

"And now I want you off my goddamn island."

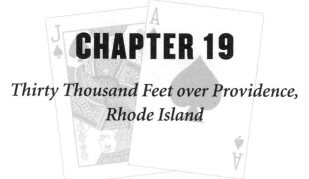

CHAPTER 19

*Thirty Thousand Feet over Providence,
Rhode Island*

emyon couldn't decide whether to be proud or ashamed.

"Banned from an entire island," Allie said, shaking her head as she stretched her long legs out into the cramped space beneath the airplane seat in front of her. "Not just kicked out of a casino, not just barred from a hotel. But banned from an entire island. That's pretty impressive."

Semyon laughed, but his head was still spinning from the ordeal. Even four hours into their flight back home to Boston, he was still going through the events in his head, reliving the absurd moments that preceded their airlift back to the world of common sense.

After they'd been forced to empty their pockets onto the golf course, Sterling had led them back to the police cars. They had been placed in one of the cars by two uniformed cops, who had locked the doors behind them. The two cops hadn't said a word to them during the short ride to the airport; no matter what they asked about—their bags, their hotel arrangements, Allie and Owen—they were met with the same stone-faced silence.

Once they'd arrived at the airport, the two cops had led them directly to the American Airlines gate. To their relief, Owen and Allie had been waiting for them by the check-in counter, along with their packed suitcases. There was also an official from the government of Aruba waiting with their stuff, a rotund, dark-skinned man who spoke with a heavy English accent. He told them, in officious tones, that they were no longer welcome in the protectorate of Aruba and that they were now officially persona non grata. If they returned, they would be in violation of numerous trespass laws and could face immediate imprisonment. Then he'd handed them each a signed document to that effect, along with airplane tickets. At least the tickets had been comped, Victor had pointed out as they'd been ushered onto the waiting plane. So the trip wasn't entirely a waste of time.

"I can't believe he pulled a gun on you," Allie continued, touching the oval window next to her with her outstretched fingers. They were somewhere over Providence, only about fifteen minutes away from beginning their descent toward Boston. "You think he really would have shot you if you'd refused to give him back his chips?"

Semyon shrugged. There was concern in Allie's voice, but also a tinge of excitement. He had to admit, there was something titillating about the high-stakes game they were playing. In a little more than a month, Owen had been handcuffed to a chair, Semyon had been beaten up and thrown out of a moving car, and they'd been robbed, at gunpoint, on a millionaire casino owner's golf course. They'd also won close to three hundred thousand dollars, been forced to return over one hundred thousand of it, and had at least one Vegas based private eye hot on their tail. Card counters had to work years to reach that level of excitement—they'd only been playing for a matter of weeks.

Maybe he was crazy, but even after he'd had guns pointed at him twice, the danger of what they were up to still seemed very unreal. It was a game, an exciting one at that. This round, they'd lose, but who knew what would happen next?

Semyon looked across the aisle and saw that both Victor and Jake had fallen asleep in their seats. Jake was sprawled out against the window, jammed so tightly into the confined area that it looked as though it would take the jaws of life to get him out. Victor was curled up in a shiny little ball, his agile features relaxed, and for the brief moment he looked innocent, like he didn't have a care in the world. Semyon shifted his eyes two rows back—and spotted Owen in the aisle seat. Owen had his tray table down, and there were three little bottles of Scotch lined up like miniature soldiers. Semyon noticed that Owen's face seemed a bit limp, and there were worry lines around his eyes. It bothered Semyon enough to get him up out of his seat.

"I'll be right back," he said to Allie as he undid his seat belt and headed down the aisle.

Owen slid over into the empty window seat as Semyon approached. Semyon lifted the tray lined with whiskey bottles high enough so that he could slide into the aisle seat. The smell of the liquor was strong enough to make him cough.

"Are we celebrating something?" he asked. "I mean, it's true, you don't get kicked off of an island every day—"

"Semyon, I fucked up."

Semyon paused, seeing that Owen's expression was dead serious. Owen was trying to peer over the seat in front of him, and it was obvious where he was looking.

"Victor's asleep," Semyon said, his voice low. "What do you mean you fucked up?"

In his mind, he could see the police record that Victor had handed him in the bathroom at the Trump Plaza. He quickly pushed the thought away. There was no way this had anything to do with that.

"After we left you guys—me and Allie. We separated, and I wandered into the other casino in the resort. I played a little blackjack on my own. Tried to do a few ace cuts."

Semyon frowned. It was against team rules to play on one's own and doing cuts alone was pretty foolish. It was hard enough to pull it off with two people, let alone one. But still, it wasn't something to get so worked up about. Unless . . .

"How much did you lose?"

Owen paused, glancing again toward Victor's seat. Then he leaned in close.

"Seventy-five."

Semyon raised his eyebrows. Seventy-five? What did that mean? Seventy-five hundred? No, they only talked in thousands. Seventy-five thousand. Owen had lost seventy-five thousand dollars.

"Are you fucking kidding me?"

Owen shook his head, then shut his eyes and leaned his head back against his seat.

"Fuck, what am I going to do? Victor's going to kill me."

Semyon took a deep breath. Victor was going to do worse than kill Owen. He was going to kick him off the team. Victor already had it in for the kid; now he had real reason behind his distrust. Shit, if Semyon had been in charge of things, what would he do? Would he have kicked Owen off for the loss? He wasn't sure. People made mistakes. People fucked up. One could make the argument that Jake's signaling them to play the second high hand was what got them caught at the casino in Aruba; too much, too fast. Was he going to get reprimanded for the move? Of course not. So should Owen get kicked off the team for one mistake? Even if it was a very costly mistake, at that?

All Semyon knew was that he didn't want to lose his friend. He fingered one of the tiny Scotch bottles, thinking. Then he sighed.

"You let me handle Victor. I'll think of something. Let's keep this between us, right now. If Victor asks, you tell him you passed me your chips at the casino, right before we got nailed. He'll never know that you got out of there with chips on you."

It was a big lie, but Semyon was hoping he'd figure some way either to pay the team back or to explain things to Victor in a way that would save Owen. If Semyon was in charge, he'd have made Owen's mistake into an example and let them all learn from it. But Semyon wasn't in charge, and he knew that Victor would simply overreact.

They had no choice but to play by Victor's rules. And by Victor's rules, a mistake as costly as Owen's would be dealt with severely. Their only real move, for the moment, was to cover it up. Semyon didn't feel good about the situation, but there wasn't much else he could do.

The captain's voice came over the PA system, signaling their initial descent into Boston. Semyon gave Owen a thumbs-up, then returned to his seat next to Allie.

"Everything okay back there?" she asked. Semyon nodded. He

had a feeling he could trust Allie, but he didn't want to bring her into this. It was between him, Owen, and Victor.

Semyon looked over at their resident guru, who was still asleep in his cramped airplane seat.

Semyon wondered what Victor would do if Allie had made the same mistake, had lost seventy-five thousand dollars of the team's money. Or what if Semyon had been the stupid one, trying to do cuts on his own? Would either of them warrant special treatment, or were they all just cogs in Victor's blackjack machine?

Semyon pushed the thought away. They weren't cogs, they were partners. The enemy wasn't within their ranks—it was the casinos, the monsters who built the game in the first place. The monsters who would bar them from an entire island simply because they were smart enough to win.

Seventy-five thousand dollars was nothing, in the greater scheme of things.

CHAPTER 20

Boston, Massachusetts

Twenty-five thousand dollars in a sock drawer.

Fifty thousand dollars in a cardboard box, jammed into the back of a cluttered closet in an MIT dorm room.

One hundred and nineteen thousand in a hollowed-out black-and-white TV set, sitting on a shelf of discarded electronics in an infrequently visited MIT lab.

Two hundred and sixty thousand in the shell of a computer processor, in the basement of a frat house on Beacon Street, just over the bridge from the main campus.

And nine hundred thousand more in laundry baskets all over the city, banded stacks of hundreds stuffed beneath dirty pairs of underwear.

Eyes closed, lying on his back on a blanket on the freshly cut grass of the Esplanade, the bright green park that runs the length of the Charles River that separates Cambridge from Boston, Semyon imagined he could see all those stacks of hundreds, their different hiding places glowing in his mind like brilliant points of light in a connected universe of wealth. He himself had ten of those stacks in a dresser drawer in his shitty little apartment in Brookline; the bills were worth more than the apartment, but of course they weren't his to spend, not yet, anyway. They were the team's money, part of the profits they'd made in the past ten weeks of play. Ten weeks of trips to Vegas, ten weeks of weekends spent at various Strip hotels under various names, playing the three techniques in shifts of three and four hours per casino. The names, as varied as the hiding places where they'd been stashing their loot, spun through his thoughts: Nikolai Nogov, the Russian gun dealer. Parker Buckingham, the trust-fund brat. Michael Nussbaum, the neurotic dentist. Dietrich Kruger, the German Euro-pop star. Miles Klinger, a computer programmer who'd just won the lottery. Charlie Jones, a divorced alcoholic. And his current personal favorite, Dicky Daze, an actual pimp, with Allie, Misty, and Carla as his stable of ready-to-please hos.

It was amazing how easy it was to fool the casinos when you had those stacks of hundred-dollar bills at your disposal. Even with Jack Galen—and his friends at Griffin—wandering around behind the scenes, they hadn't run into any serious heat since Aruba. The hardest part of the game, actually, had been dealing with all the cash, transporting it back and forth from Vegas, stashing it in hotel rooms, in luggage, in tiny little hotel safes, and stowing it away back in Boston. Using bank accounts didn't make sense for a variety of reasons: First, because they were hitting Vegas so often, they needed large amounts of cash to be liquid and available, and withdrawing a million bucks one week to deposit it the next was more than a little suspicious, even if what they were doing was completely legal. Keeping bank accounts in Vegas was likewise impossible, since they'd have to use real IDs to get their cash, which would leave a paper trail that would be easy for Galen and his ilk to follow. And they knew how moving around such large sums of cash would pique the interest of the three letter organizations: IRS, DEA, FBI, etc.

So for the moment, they'd invented their own form of "money laundering," literally keeping the large bulk of their cash hidden in their laundry. At some point, the profits—now around six hundred and fifty thousand dollars—would be counted and disbursed among them. Since Victor had made the large initial investment, had recruited and trained them all, he'd be getting the lion's share. The rest of the cash would be divided among the other six, based on playing time. Semyon calculated that he'd earned about seventy grand already, which wasn't bad for ten weeks of play. If he and Owen were to come clean about the seventy-five thousand Owen had lost, and Victor made them pay it back out of pocket, those earnings would be cut in half—but Semyon hadn't come to a decision about that yet. The only thing he had decided was that he was going to share the loss with Owen, either way, because he had helped cover it up after the Aruba trip. If Owen was off the team, Semyon would leave as well. But he hoped it never came to that. He was having too much goddamn fun.

He opened his eyes and stared up at the blue noon sky. A pleasant wind was blowing in over the Charles, tugging at his MIT sweatshirt. It was a perfect Monday afternoon. He was still getting over the late-night flight back from Vegas of the night before, and he intended to spend most of the day lying by the Charles, watching the pretty young MIT summer-school coeds jogging by.

His plan went to hell the minute the folded newspaper landed on his chest.

He sat up and saw Owen strolling toward him, ambling down the center of the paved path that ran parallel to the river. Two girls on Rollerblades, in spandex shorts and bikinis, blew past, momentarily catching Owen's attention. When his eyes met Semyon's, he said, "You're not going to believe this."

Semyon unfolded the newspaper and looked over the front page. It was a copy of the *Boston Globe*, that morning's edition.

"What am I looking for?"

"At the bottom," Owen said, dropping onto the grass next to him. He was still staring after the two girls on Rollerblades, his eyes bouncing in tune with their spandex-covered rears. "The last column on the right."

Semyon saw the headline and his eyes went wide.

JANITOR FINDS $150,000 IN MIT CLASSROOM.

Semyon quickly devoured the small article. A janitor cleaning up an MIT classroom along the Infinite Corridor had found a shopping bag beneath one of the desks. He'd been about to throw the bag out when he'd looked inside and discovered fifteen stacks of banded hundred-dollar bills. He'd turned the bag over to the school's lost and found, and now the administration had called in the DEA, assuming it had something to do with drugs.

"Um," Semyon said.

"Yes, um," Owen responded. Semyon put the paper on the grass between them.

"I assume it's our money?"

Owen shrugged. "Victor headed over to the classroom last night, after we landed, to go over our notes. He had more than half of our stash with him."

Every trip, they kept detailed notes that documented their playtime, expected earnings, and actual win-and-loss rate. After each trip, Victor went over the notes in meticulous detail, to keep track of their progress. Semyon had a suspicion that Victor also went over the notes to see if there were any discrepancies between what they were expected to win and what they had actually won—but if he had found any problems so far, he hadn't mentioned it. Semyon trusted everyone on the team, and he assumed—despite Victor's problems with Owen—that Victor did, too. There had to be trust, otherwise the team wouldn't be able to function.

"So Victor fucked up," Semyon said.

He saw that Owen was smiling. It was true, it made Owen's seventy-five-thousand-dollar loss seem much less significant. At least he'd been playing blackjack. Victor had left a bag full of cash in a classroom. Now that was something to explain.

"I assume Victor's called a meeting to go over this?" he asked.

"Actually," Owen said, glancing out across the Charles toward the clearly visible white-domed building that marked the center of the MIT campus, "Jake called the meeting, not Victor. Although it was Victor's fuckup, Jake says he has a solution. We're supposed to meet on campus in an hour."

Semyon raised an eyebrow. "On campus? What, we're going to return to the scene of the crime?"

"Not the classroom," Owen responded, running a hand through his wild locks of hair. "Someplace else. Someplace, well—kind of strange."

The building was low and beige and boxy, squatting at the end of a narrow alley that led from one of the campus's many physics buildings to an obscure engineering library that Semyon recognized from his first week on campus. He'd gone to the library in search of a book on a certain form of digital wiring, but he'd never been anywhere near this alley, had never even glanced in the direction of the beige building with the clouded windows and the nondescript, unnumbered steel-paneled door. The building looked pretty much abandoned, which was interesting, considering the proximity of the real estate to the Charles River and the main center of the MIT campus; interesting, though not particularly surprising, since the campus had been going through the convulsions of a major architectural redesign for decades now, and many of the older buildings sat unused, waiting for the administration to figure out what to do with them.

By the time Semyon and Owen arrived at the end of the shadowed alley, the rest of the team had already gathered. Victor was standing at the front of the small group, a sheepish look on his face. He looked up as Semyon and Owen approached.

"Well, guys, I'm sure you've heard about my little indiscretion. I've already contacted my lawyer, and he's assured me he can get the money back. But it's going to take a while, and cost a bit, which I'll cover from my share."

Semyon wondered if it would really be possible to get the money back. He doubted that Victor had any way to prove that the cash was theirs, and did they really want to publicize the fact that there was a cabal of professional cardplayers using an MIT classroom as a training and recruitment office?

But he decided to keep his mouth shut as Jake took center stage, moving to the steel-paneled door. He had a large metal skeleton key in his hand. With a deft twist, he unlocked the door and yanked it open. Semyon saw an unlit, cement-floored hallway and a set of steep stairs, leading downward.

"A bit of history, boys and girls. This building is about a hundred years old, one of the oldest on campus. Until about ten years ago, it was in continuous use as a storage facility for the nearby physics lab. During World War Two, they actually kept some of the earliest components of radar in here, and you had to have some serious clearance to get past this steel door."

Allie was tightening the strings of her sweatpants as she listened. She looked like she'd just come from the gym. "How do you know all this, Jake?"

Jake grinned. "I did some work last summer as a paralegal for the university. This was one of the buildings earmarked for demolition during the last campus reengineering phase, but it got left off the list because nobody could decide whether it had historical value or not. Anyway, I had to check it out a few times, and that's how I made my little discovery. If you'd all care to join me, I've got something pretty incredible to show you."

He gestured toward the open doorway, inviting the rest of the team to follow him inside.

It was, in a word, magnificent.

A perfect cube, more than seven feet high, taking up an entire corner of the dank, cement-floored basement. Thick steel walls, an even thicker steel door, now hanging wide open from massive iron hinges. The interior of the thing was about the size of an industrial freezer, and on the back of the open door, there were polished steel levers, gears, and tumblers.

The safe was so big and ominous that the rest of the basement seemed cramped and dingy, even though the place was at least thirty yards across and pretty well lit by crisscrossing fluorescent tubes attached to the cement ceiling. There were shelves lining two walls, cluttered with the remains of machines too old to identify, a sort of electronics museum that probably would have made Semyon's dad blush with excitement.

Jake took a step forward, touching the massive safe's door with his thick fingers.

"The university used to keep the money they used for payrolls in here before World War Two. This safe weighs more than seven tons. Its walls are reinforced steel, so thick you'd need dynamite to get through it—and even then, I'm not sure you'd be able to crack this thing. It's a hell of a monster, and it's big enough to fit our entire stash in between trips."

Victor stepped forward, taking Jake's place, and put his own delicate hands on the steel door. He gave it a shove, and the door swung shut. The sound reverberated through the basement.

"I've already reverse-engineered the combination," he said, pointing to the massive, circular, numbered wheel in the center of the safe door. "It took about six hours, actually. Jake and I've been planning this for a few weeks, and the events of this morning have sealed the decision for us."

Semyon looked around the basement at the cobwebbed shelves. If he was hearing Victor and Jake correctly, they were planning to use this safe, in the middle of the MIT campus, to store their money. He was impressed that Victor had had the engineering ability to figure out the combination—something he was pretty sure he could have done himself, but he'd never pegged Victor for such a mechanical geek—and even more impressed by the sheer hubris of the idea.

"What if someone else has the combination?" he asked.

"Come on," Jake responded. "This safe is like seventy years old. It's from before World War Two. Anyone else who knew the combination is long dead."

There was a brief pause. It was Misty who finally broke the silence, clapping her hands. "I love it. It's so fucking crazy. We keep the money right here on campus."

Semyon had to admit, the idea had a certain flair to it. It was dangerous—if the administration got wind of the situation, God only knew how they'd react. But it wasn't any more dangerous than keeping all that money in their laundry.

Jake stepped forward again, next to Victor, and leaned his huge bulk back against the safe door. Of course, the thing didn't budge. Then he held out a hand toward Carla, asking her to step forward. When she reached his side, he put his thick arm around her shoulder. He had another surprise for the team, and it had nothing to do with blackjack or seventy-year-old safes.

"I've also brought you all here to make a little announcement. Carla and I are getting married."

The pronouncement had come out of nowhere, and it took a moment for the rest of the team to react. Then Misty and Allie jumped forward, giving Carla a double hug. Semyon, Victor, and Owen congratulated Jake. It was a surprise, but Semyon had seen it coming for a while. Even though Jake and Carla seemed young to be getting married, Semyon knew how fast relationships could grow under the heat of a team that spent so much time together. Being on a blackjack team was like being in a pressure cooker; everything seemed accelerated.

As he stepped back from the celebrating crew, he couldn't help glancing toward Allie. She was still hugging Carla, and her head was turned away, but he could see her profile, the high ridge of her cheekbone, the pretty upward angle of her full lips. He felt warmth in his chest, and he had the sudden urge to pull her aside, to tell her that something was brewing inside of him, something he wanted to test in real life. But before he could even really contemplate the action and toss it aside, he felt Victor's hand on his arm, drawing him farther away from the rest of the group.

"Semyon," Victor said quietly. "If you've got an hour to spare, there's something else I want to show you."

CHAPTER 21

Hanscom Airfield
Outside Boston, Massachusetts

To Semyon's surprise, the thing Victor wanted to show him wasn't located anywhere near the MIT campus. In fact, it took them a good thirty minutes to drive the twenty miles northwest of Boston—in Victor's cramped Toyota Corolla, which he'd borrowed from his present girlfriend, a Tufts grad student none of them had met, or would meet, considering she'd probably be gone by the end of the month—to a place Semyon had never heard of before, much less ever visited.

Hanscom Airfield was a class operation, though compared with a real airport, it seemed very bare bones and more than a bit terrifying. The place consisted of a number of airplane hangars surrounding two runways that served a variety of civilian, business, and even military planes.

At the moment, Semyon and Victor were standing at the edge of one of the smaller hangars, a domed, prefab construct that contained a line of what looked to be small two- and four-seater civilian planes. Semyon didn't know much about airplanes and had never been in a plane as small as the ones he was facing. He was pretty sure that he didn't want to break that streak. The things looked flimsy, like a strong wind could bust them into pieces.

"So what the hell are we doing here?" he asked. He'd much rather have been out celebrating Jake and Carla's engagement with Allie and the rest of the team. Although he and Victor had spent plenty of time together in the past ten weeks of blackjack play, they didn't have much to talk about outside of the game. Semyon wasn't even sure he'd have described them as friends, just associates. He didn't dislike Victor, but ever since Atlantic City, he hadn't tried to get too close to the young man.

Victor smiled at him and pointed toward the closest plane. It was a bright yellow Cessna, and from Semyon's vantage point, it looked to be in pretty poor shape. In fact, it looked more like a VW Bug with wings than a working airplane.

"That's what you wanted to show me?" he asked. "An airplane?"

"My airplane," Victor responded. His smile was wide and a bit feral. "I bought it two weeks ago. I've been taking flying lessons for the past month. I'm going to be ready and rated to fly by next weekend."

Semyon stared at him. Then he turned back toward the little airplane. No fucking way, he thought to himself, am I getting into that death trap.

"Have a nice trip," he finally said.

Victor laughed, slapping Semyon's shoulder.

"Come on, man. Don't you see? This is going to save us tons of money. We can hit Atlantic City at will, carry our money without worrying about airport security. It's perfectly safe."

Semyon rubbed his jaw, looking at the chipped yellow paint that was peeling, in jagged triangles, from the wings.

"Why did you bring me here, instead of the whole team?" he asked.

Victor sighed. "If you haven't noticed, Semyon, I've singled you out from the rest. I think you're the best of the group—as good as me, actually. I think of you as my partner."

Semyon nodded.

"We're all partners," he said, though of course he didn't really think of it that way. He and Owen had been calling Victor their boss since Aruba, and he knew that Allie, Carla, Misty, and even Jake felt the same way. Teammates, yes, but not partners.

Then Victor tossed him another surprise.

"From now on, I want to give you an even share of the profits."

At first, Semyon wasn't sure he'd heard Victor correctly. An even share of the profits? Depending on how much they made, that could work out to hundreds of thousands of dollars—maybe even more.

"Are you serious?"

"I can't do this alone," Victor said. "If we keep going the way we're going, we're going to be talking about millions, not hundreds of thousands. I need a partner I can trust—and one who trusts me."

Semyon took a deep breath. The seventy-five thousand dollars that Owen had lost flashed through his thoughts. He wanted to bring it out, right now, but it just didn't seem to be the proper time. What Victor was offering him was, well, immense. If he was reading Victor correctly, he'd be co-running the team. He'd always assumed that Jake was going to end up Victor's partner, in charge of things, but then again, maybe he was the better choice. Jake wasn't from MIT, and his knowledge of math lagged behind the rest. And now that he was marrying Carla, he'd probably be spending less time as a player. The blackjack team was a pressure cooker, and that could work against relationships as easily as it could work for them.

Obviously, Victor had thought this through. Now it was Semyon's turn to give it some thought.

"It's an incredible offer," he said finally.

Victor grinned at him, then put his arm on Semyon's shoulder, aiming him back toward the crappy little yellow airplane.

"You do trust me, don't you, Semyon?"

CHAPTER 22

Midtown Manhattan

PRESENT DAY

The woman was short and muscular and poking at my face with an eyeliner pencil. I wanted to dodge the jabs, but I knew that it would just drag this out, and I'd already been sitting in the makeup chair for ten minutes. I was pretty sure I looked the same as when I'd first sat down, but I wasn't an expert in the fine art of applying makeup for daytime television.

It was a little after seven in the morning, and I was seated in a room about the size of a New York City studio apartment—which meant I could almost reach all four walls without ever leaving my chair. There was a large rectangular mirror in front of me, backlit by enough candlepower to burn a hole in the ozone layer. Aside from the woman working me over with the eyeliner pencil, there was one other person in the room, another guest sitting in a chair similar to mine, facing the same brightly lit mirror, reading a tabloidy magazine. He'd already had his fun with the makeup girl and was just waiting for the executive producer to return from wherever it was she had gone, to tell us that it was time to go on air.

It was an interesting, well-choreographed world, backstage at a daytime talk show. Much of it was a waiting game; hours spent sequestered in tiny green rooms or makeup studios, being prodded and pruned by women who usually smelled like cigarettes and always made small talk about their damaged love lives. Often, when the women finally left you alone, you were forced to make conversation with the other guests sharing your time in purgatory; usually people who had little to say to you, and even less interest in saying it.

Backstage at the *Today* show, I once had a ten-minute conversation with Charles Barkley about the best barbecue restaurants in New York. Outside the Catherine Crier show at Court TV, I spent twenty minutes in a green room with the former head of the Bonanno crime family, arguing about which was our favorite Cirque de Soleil show in Las Vegas. At CNN, I'd been relegated to the group green room they used for the way-below-B-list celebrities—and of course authors—where I'd shared little paper cups of water with an elderly, married couple who had made millions producing porn movies. Yes, doing daytime TV was a glamorous thing.

As a writer, you do things like this. You go on talk shows and smile and tell jokes and beg the people on the other side of the camera lens to buy your book. You wait around green rooms and make conversation with people you'd normally run away from, because this is part of your job, and actually, in a lot of ways, the better part of your job. Because anything beats sitting at your desk and actually writing.

Except this morning, in the makeup room of *The Montel Williams Show,* the guy sitting next to me happened to be someone I actually wanted to talk to. His name was Martin Leonard, and he was one of the premier Las Vegas casino hosts, a man who'd worked for nearly every casino on the Strip at one time or another. Like most hosts, he was well presented and preserved: tan, handsomely weathered face, a crown of salt-and-pepper hair, and a suit that looked like it had been pressed five minutes ago.

It wasn't merely a coincidence that Leonard and I were sitting next to each other. We had both been booked to talk about gambling—or more accurately, to instruct teenagers on the dangers of gambling. I wasn't entirely sure that I was someone who really should be lecturing kids on the dangers of gambling, as I'd made a mint in the past couple of years telling stories about kids who gambled, but I'd been booked just the same. Leonard was an even odder choice for the show, considering he actually made his living *encour-*

aging people to gamble; but nobody could ever blame Montel Williams for putting together a boring television show. In any event, Montel's choice of guests had given me a perfect opportunity to write off my trip to New York as a research expense.

I waited until the makeup woman finished with my eyeliner and left the room before I introduced myself to Leonard. He recognized my name from my book jacket and didn't seem to mind when I jumped right into my line of questioning.

"As a host, does it matter to you where your clients get their money?"

I pretty much knew the answer before I asked it, but sometimes it was better to get to the good stuff by taking a roundabout route. Leonard folded up his magazine and smiled at me in the mirror.

"We just care about the bottom line—what sort of gambler they are, how much our whales are willing—and able—to bet, and what it's going to take to get them into our casino."

I nodded. Even I'd used hosts before, as a fairly frequent Vegas visitor, and most of them had never asked me about my day job. They'd just wanted to watch me gamble and see what sort of comps I was capable of earning. But when someone like Semyon Dukach had hit Vegas, it was on a whole different scale. And it seemed amazing that a twentysomething kid could walk into a casino with half a million bucks in his pocket and not raise a lot of questions.

"And if your client turns out to be, well, not so legit—"

"That's really not my problem," Leonard interrupted. "I'm not a cop, I'm a casino host. Hell, I don't even care if one of my high rollers is actually a card counter, or a cheater; that's for casino security to figure out. I just bring in the big-money players. The more action I can bring to the tables, the more I get paid. It's a pretty simple game."

I decided to change tack, direct the conversation more toward what I was really trying to find out.

"What sort of players are you most interested in?"

He shrugged. "I've had guys who would gamble a million dollars a night. Others were hundred-thousand-dollar players, not even whales, more like fish, guppies. But it's the million-dollar players that really make the difference."

The million-dollar players—the true whales. The people Semyon and his team were trying to pose as every time they walked into a casino.

I decided to take my questions one step further.

"I'm curious if you've ever heard of a friend of mine. He was a pretty big gambler about ten, fifteen years ago. Went under a few different names when he was out in Vegas."

"If he gambled big, I'm sure I knew him."

I tried a few of Semyon's aliases, looking for a hit. Mike Nussbaum. Parker Buckingham . . . Nikolai Nogov—

A genuine smile broke out across Leonard's face.

"Yeah, everyone's heard of Nikolai Nogov. The Russian. Some sort of businessman. Something dirty, maybe guns. Nogov was the real deal. A whale, and a big one. One of the biggest. Actually, he was so big we even had a name for him."

Leonard straightened the lapels of his jacket, winking at me in the mirror.

"The Darling of Las Vegas."

I nodded. I'd heard the nickname before, first from Semyon himself, then from a few of my sources in Vegas and elsewhere. It was a real catchy title, one that I was sure Semyon had enjoyed. *The Darling of Las Vegas.* And it was especially ironic, considering that the darling had actually been a twenty-two-year-old kid who had been busting Vegas with a system he'd learned in an MIT classroom.

"Every host wanted a piece of him," Leonard continued. "We all wanted his action. A guy like that could make a host's career. If you could bring him into your casino, then you could basically write your own ticket. Yeah, I knew all about Nogov."

I wasn't exactly surprised, but the excitement in his voice did stun me a bit. Fifteen years later, and a casino host still remembered Semyon so vividly. He knew him only as Nogov—but I could guess that other hosts fondly remembered other of Dukach's aliases.

"So you worked with him?" I asked.

Leonard shook his head. "No, but I knew about him. I knew that he was one of the biggest high rollers the city had seen in years. I knew he always carried cash, big cash, and that he traveled with a blond girl who could turn a head from a hundred yards."

He glanced behind him, toward the door to the makeup room, then back at the mirror.

"And I also know that one day, he just disappeared."

I raised my eyebrows. I could see, from the look on Leonard's face, that the statement was meant to sound ominous.

"Maybe he just decided to quit," I said. "To stop gambling."

A production assistant stuck her head into the green room, said something about us having five minutes until we were going on. When she shut the door behind her, Leonard leaned toward me.

"These guys, they don't just stop. Oh, some of them go broke. They play until there's nothing left, until we basically have to kick them out. But they never give it up on their own. They just don't. It's not in their nature."

He leaned back, checking his crown of marbled hair in the mirror behind my head.

"If they disappear, it's usually because something happened. Something bad."

I breathed deeply. I could smell the makeup on my own face, and it struck me that I was about to go on TV, I was about to tell teenagers how awful gambling could be, how dangerous it could be to get caught up in that world. The funny thing was, the conversation that I was having right now—which we would never have dared to have on television—would have been more of a warning than anything I would soon be saying on *Montel*.

"The Darling of Las Vegas," Leonard continued. "He just disappeared. He hit the town like a windstorm, and then just—disappeared. Like a fucking ghost. Yeah, that's what he was. Some sort of ghost."

Semyon Dukach, a ghost who had survived Aruba, Atlantic City, and a plane crash that nearly took his life—and was still one of the biggest legends the neon city had ever known.

I realized that Leonard was looking at me in the mirror.

"Hey, if you ever run into him, you give him my card, okay?"

Even as a ghost, Semyon Dukach was still the Darling of Las Vegas.

CHAPTER 23

Boston, Massachusetts

The thing about prescription painkillers—the really good ones—is that if you take enough of them, it's kind of like entering a time machine. The rest of the world keeps on moving forward, but you're stuck in one place, lodged in a timeless fog. You have a chance to think, but really you can't, your brain's all fucked up, your synapses aren't quite lined up anymore. You go through recent events in your head, but it's kind of like watching a movie through dirty glasses; there's no chance of achieving any clarity.

Eyes closed, lying flat on his back in his bed, Semyon relived the plane crash once again, as he had countless times in the three weeks since he'd been released from the hospital. He watched that line of planes coming closer and closer, experienced that brick wall that suddenly seemed to leap up and catch the landing gear, the fireball that engulfed the cockpit, the wings curling inward, the noise, the heat, the searing heat. He watched himself diving out into the cool night air, running away from the explosions and the fire and the heat—and then he saw Victor, crazy fucking Victor, heading back toward the plane. Back to retrieve the garbage bag full of money.

He should have just stood there and let Victor do what Victor had to do. He should have just waited for the ambulances and the police cars to arrive. Instead, he went right in after the crazy bastard, he dove headfirst into a burning airplane. Together, they pulled the bag of money free. Like Victor had said, they were equal partners—and in that moment, they were equally insane.

After they'd pulled the money from the plane, and Semyon had collapsed back onto the grass, the rest of the fateful evening sped by, more of a blur than a spinning movie reel; flashed images more like sequences from a dream than anything resembling real life. The frantic trip to the hospital, strapped to a gurney in the back of an ambulance. Victor clutching the bag of cash as they put him in a wheelchair, even though he didn't really need one—somehow, he had avoided any injuries beyond a few second-degree burns on his neck and back. Semyon had fared much worse; his right foot had required surgery to realign the bones, and his hands had been burned bad enough to require a visit from a plastic surgeon. Luckily, the surgeon had eventually decided that the burns, for the most part, would heal on their own, and a skin graft had been avoided.

Three weeks later, most of Semyon's wounds had indeed healed. His foot, still wrapped in bandages, had finally stopped hurting, though he still needed a cane to get around. And the scars on his hands had almost entirely faded. The thing that hadn't healed, however, were his strained feelings toward Victor—because he couldn't help but blame what had happened on his partner. If Victor hadn't been so cheap, they'd never have gotten inside that ridiculous excuse for an airplane. And if Victor hadn't been so fucking proud, he never would have thought that he could fly the thing at night, on his very first try.

But Semyon had, for the most part, hidden his anger from Victor and the rest of the team. They'd all been so kind to him during the healing process—bringing him food, taking care of him, even washing his clothes and cleaning up around his apartment—that he didn't want to rock the boat. Allie, especially, had been overly attentive toward him, and had even spent the night a few times—though of course, she'd slept in the living room, leaving him alone and bandaged in the bed. He felt pretty certain that there was something brewing between them, though neither of them had mentioned it, and there hadn't yet been any real physical signs. The last thing he wanted to do was strain his relationship with the team by revealing his anger toward Victor. Allie still saw Victor as the guru who had given them all the gift of blackjack; no doubt, she was under his sway. It wouldn't do Semyon any good to admit his own feelings.

Instead, he had simply dedicated himself to the job of healing and had spent most of his time lying flat on that bed. When he had the energy, he practiced his cuts and his sequencing, and he was pretty sure he was now as good as he was ever going to get. He was actually getting quite eager to return to a casino—the longer he spent in his room, the more he felt the pull of Vegas, the call of the blackjack felts.

Opening his eyes, staring up at the blank ceiling of his apartment, he knew that he was ready to get back into the game. More wary, now, of Victor than ever before—but still a partner and a player, his skills more honed than ever. He was ready to play.

He sat up—and felt something slide down his chest, something the size and shape of an envelope. Somebody had obviously put it there while he'd slept off his last few painkillers. The only person who had a key to his apartment was Allie, so he assumed it was from her.

He found the envelope and tore open the seal. Inside, he found an invitation, elegantly and professionally done, with gold script and raised lettering. He smiled as he looked the thing over—at the sheer audacity of it all.

The invitation was to a wedding, at a hotel and casino in Las Vegas. Jake and Carla's wedding, to be specific. And not just any casino—it was the one with the infamous back room, steel chair, and handcuffs. From the looks of the invitation, it was going to be quite an affair. Black tie, free accommodations, even airfare for the guests who needed it. Victor, the best man, had added a little note at the bottom of the invitation, thanking the hotel for putting together this special occasion.

Semyon shook his head, absolutely awed.

Jake and Victor had managed to get an entire wedding comped, down to the goddamn invitations.

The suite was almost four thousand square feet, two levels with massive picture windows that overlooked the Strip, marble floors, walls, and ceilings, and more chandeliers dripping teardrops of crystal than Semyon had ever seen in one place in his life. There were two winding staircases at either end of the circular living room, each curling upward like two parts of the same Möbius strip. Both of the staircases had been lined with thick red carpeting, just for the occasion. The uniformed waiters had already handed out the champagne, and the forty or so guests waited in silent anticipation as the music from the violin quartet—squirreled away in some alcove on the second floor—echoed through the cavernous suite.

Semyon was leaning back against the wall closest to the door, his cane braced up against his right leg to take some of the pressure off his aching foot. It had been difficult getting the leg of his rented tuxedo pants over the brace that enveloped his ankle, and he hadn't even tried to do anything to disguise the white wrap that covered it. Otherwise, he thought he looked pretty good.

On the other hand, Allie, standing next to him, looked absolutely amazing. Her silver dress, tight as a second skin, hugged her curves with such precision Semyon was having trouble looking at her directly. And when she moved, the shimmering silver separated just enough to reveal a sliver of porcelain thigh. Nikolai Nogov would have been proud to parade her through the casino—though of course, Nikolai Nogov wouldn't be setting foot anywhere near the blackjack tables of this particular casino.

Semyon had been more than a little apprehensive about returning to the same casino where he and Owen had been nabbed. Even though Victor had assured him that enough time had passed to cool down the heat on them, that if they were careful and avoided the gambling areas, they'd be safe, Semyon had seriously contemplated skipping the wedding altogether. Only the thought of Allie, dressed up and drinking champagne, had made him change his mind. And if he was apprehensive about a chance run-in with Jack Galen again, he couldn't imagine what Owen was feeling, considering his last visit to the casino had landed him in handcuffs.

At the moment, Owen didn't look concerned, that was for sure. Standing to Allie's left, his arm around the waist of a tall, blond woman with enormous fake breasts, he seemed as happy as he'd ever been. The woman was certainly attractive, if overly tanned and showing way too much skin—even for Vegas—and hadn't left Owen's side since they'd congregated at the suite. Allie was convinced she was a stripper from one of the local clubs; Owen had introduced her as Kimberly, no last name, which supported her theory, but Semyon hadn't had a chance to press the matter, because minutes after they'd met up in the suite, the waiters had informed them of the bride and groom's impending arrival.

As the music neared a crescendo, Semyon shifted his eyes back toward the closest of the two curved staircases. He had to crane his neck to get a good view of the red carpet through the impressive throng of people; he didn't know most of them, though he recognized more than a few from the card-counting team that Victor and Jake had been recruiting and training during his first few weeks with the team. A few of the faces were also familiar from the Infinite Corridor, and from physics and math labs on campus. He wasn't really sure how big Victor's card-counting endeavor had grown, but from the looks of the room, he had quite a few disciples outside of their own little team.

Just as the music hit its final notes, Semyon caught sight of Victor himself, at the base of the staircase. From that angle, he could see the strip of white gauze bandaging on the back of Victor's neck, peeking out from beneath the stiff collar of his Armani tuxedo. Semyon had pretty much avoided all conversation with Victor since they'd arrived in Vegas, but he figured the gauze accessory that both he and Victor shared said enough. He wasn't sure how their relationship was going to play out over the next few days, but he intended to make the best of things. This was a wedding, after all.

As if to acknowledge the thought, loud applause broke out from the crowd as Jake and Carla strolled down the staircase. Semyon couldn't help but grin as he looked at the odd couple; Jake's huge, tuxedo-clad form dwarfed her busty little body, and she had to take the steps two at a time—hands full with her billowing white wedding dress—just to keep up with his elongated gait. But their smiles were genuine, and it was obvious they had more in common than was obvious, given their physical differences. The elegant wedding ceremony, which had taken place earlier in one of the casino's many chapels, had itself further emphasized the point; love, like math, was blind. Sometimes you had to close your eyes, forget about appearances, and just trust the numbers.

As Jake and Carla reached the bottom step, the crowd engulfed them. Champagne glasses clinked and a cheer rose up. Semyon balanced his cane between his knees so that he could toast with Allie, Owen, and the stripper chick; then he downed his champagne in one gulp. He was done with the painkillers, which meant it was alcohol's turn to take up the slack.

A waiter took the empty glasses from them, and the music changed from violins to hip-hop, filtered through a dozen hidden speakers set in the marble walls. Semyon tried not to notice how Allie's hips shifted unconsciously with the bass as Owen leaned close to his ear.

"Jake told me that they've roped off a section of the pool area downstairs for the after-party. Four Jacuzzis, three heated pools, and an open bar. I think things are gonna get pretty crazy. You want to head down with me?"

Owen winked toward his date, and Semyon had to smile. Then he glanced back at Allie, and the champagne seemed to be bubbling right through his veins.

"I'll meet you there," he said to Owen. "Save room in one of the Jacuzzis for me."

Owen nodded, took his date by the hand, and moved toward the door. Semyon took a step closer to Allie.

"Hey, Allie," he said, noticing that the words were coming a little faster than he'd hoped. "Listen, I just want to thank you for everything you did for me when I was hurt. Looking after me, and all. I really appreciate it."

Allie squeezed his left hand. He had to grab tight to his cane with his other hand to keep from falling down at her touch.

"I know you'd have done the same for me," she said. "And I kind of liked seeing you all helpless like that. I mean, we're all such control freaks, right? To tell you the truth, taking care of you kind of turned me on."

Semyon wasn't sure if she was drunk or kidding. But just to hear her say it sent fire through his chest. She smiled, playfully messing up his hair.

"Semyon, are you blushing?"

She was definitely a little drunk. She was never that physical with him—except when they were playing roles, down in the casino. Except, this was a role she'd never played. This wasn't Nikolai Nogov's bitch. This was all Allie.

Semyon finally found his voice.

"It's probably just the burns. Residual flushing."

Allie laughed. Then she seemed about to say something—but decided against it. Instead, she gave his hand another squeeze.

"I'm going to go get a quick cry in with Carla and Misty. Did I hear Owen say something about a pool party?"

"That's right."

"Well," she said, "I'll meet you guys down there. I'm not sure this dress is waterproof, but it might be fun to find out."

Now Semyon was really blushing. Before he could get his mouth working again for a response, she had disappeared into the crowd.

His head whirling, he hobbled quickly to the door. Someone held it open for him, and a second later, he was in the hallway, heading toward the elevators. There was no sign of Owen and the stripper, so he assumed they were already on their way down. He reached the bank of elevators and stood there, contemplating Allie's words. There was definitely something there. Something incredible.

One of the elevators whiffed open, and he maneuvered himself inside. Then he hit the button for the pool level, floor one. The doors shut, and he balanced himself against his cane, watching the glowing numbers as they descended. Twenty, eighteen, sixteen. The elevator was moving fast, which was helping the champagne as it journeyed through his system. Twelve, eleven, ten. *Taking care of you kind of turned me on.* Had he really heard her right? Had she really said that? And was she serious? Five, four, three—

And suddenly the elevator slid to a stop on two, one floor above the pool level. The doors slid open, and Semyon's jaw went slack, his body frozen in place.

Jack Galen was standing in front him, a friendly smile on his colorless lips.

CHAPTER 24

A Casino on the Strip
Las Vegas, Nevada

Well if it's not the Darling of Las Vegas. Fancy meeting you here, Dr. Nussbaum. Or is it Nikolai Nogov? It's hard to keep track."

Semyon didn't move. He wasn't foolish enough to think that the meeting was by any means a coincidence. Galen had obviously caught him on one of the thousands of cameras that were hidden all over the casino. Maybe in one of the hotel hallways on the way to the suite. Maybe in the lobby, even though he had kept his head down, his cane-assisted gait as quick as possible. Maybe even in this very elevator, just minutes ago. No matter where or when Galen had seen him, obviously the man had been ready, and waiting. What's worse, he'd obviously connected two of Semyon's aliases—maybe more. How far was he from figuring out who Semyon really was? And what would he do once he had that information?

"Looks like you got yourself a bit banged up," Galen continued. His tone was pleasant, but his narrow features had more animosity in them than the last time they'd come face-to-face. "Dentistry's a pretty dangerous game, eh, Doctor?"

Semyon tried to match the man's narrow blue eyes, but felt himself unintentionally wilting beneath his gaze. He found his attention shifting to Galen's perfectly tailored gray suit. Someone was obviously paying him plenty to keep tabs on suspicious high rollers. For some reason, at least that thought cheered Semyon. He and his team were obviously hurting the casinos, if they were going to pay a dandy shit like Galen enough to stay swathed in such expensive European threads.

"Yeah," Semyon answered. "Some of those extractions can be pretty hairy."

Galen ran a hand through his fancy blond locks.

"Interesting friends you've got. As you can imagine, once I saw your face on our cameras, I began collecting faces from all our security systems. Of course I found your partner from last time. I'm surprised that after what he went through last time he visited this establishment, he'd be back, but . . . well, some people never learn."

Semyon wanted to hit the elevator button, get the hell out of there. He didn't know whether or not Galen would try to stop him. After all, he hadn't been playing blackjack, he hadn't been gambling at all. The casino would have had a hard time causing him any problems for attending a wedding in a hotel suite. But despite how unnerved he was by the situation, Semyon wanted to hear whatever it was that Galen obviously wanted to tell him. *The Darling of Las Vegas.* Semyon wasn't sure where the nickname came from, but he couldn't help but enjoy hearing himself referred to that way. He was certainly making a name for himself as a high roller, even if he was catching some serious heat along the way.

"We're guests at a wedding," Semyon said, trying to sound nonchalant. "It just happens to be here."

Galen nodded. "Right, the wedding. Comped to a Dr. Michael Chavez. Supposedly some fancy plastic surgeon from L.A. Except I've checked the medical registries for all of California, and though there are plenty of Chavezes, even a couple of plastic surgeons, none of them resemble the photo we have on file of the Chavez who's garnered the comps at this casino."

Semyon shrugged.

"I'm just a passing acquaintance of the bride, and thought I'd stop by for the festivities."

Galen grinned. "A good story, except that I've got photos of you and Dr. Chavez from Aruba, taken around six weeks ago. You remember your little vacation, don't you? I was glad to be able to help out my friend down there in the Caribbean. I hear you all had a nice golfing excursion, to work things out."

Semyon wanted to take his cane and wipe that grin off Galen's face. Of course Galen had pictures of him and Victor together from Aruba, Semyon had been stupid not to assume so. But that was all he had—an association—Semyon, Victor, Owen, and Jake. He didn't know any more about their three techniques than he did when he'd first accosted Semyon. And if Galen knew they were from MIT, they wouldn't have been having this conversation here. They would already have had it in Cambridge.

"This is all really quite fascinating," Semyon finally said. "But I've had a long night, and frankly, I'm getting a little bored of our one-sided conversations."

He reached out and hit the button for the first floor. The elevator doors started to close, but Galen's hand shot out, stopping them midway.

"I want to tell you a story. There was a young man who used to come here, about three years ago. He was a high roller like yourself. Used to bet big at blackjack—we're talking ten, twenty thousand dollars a shot. And he always seemed to win. In three weeks, he was up nearly half a million. An amazing run."

Semyon could feel those narrow blue eyes boring into him, and suddenly the elevator felt like it was shrinking around him. He wanted to get out of there, but he had no choice but to stand and listen.

"At first we thought he was counting," Galen continued. "But his gains were too big, too fast. So we figured he was cheating. We followed him day and night—and finally figured out his game. Simple, really. He was marking the face cards with infinitesimal amounts of red dye. Had specially made contact lenses that allowed him to see the dye so he'd know when the big cards were coming. Really brilliant, wouldn't you say?"

Semyon shrugged. Galen tapped his manicured fingers against the elevator door.

"But as brilliant as he was, he was cheating the casino. Do you know what happened to him?"

Semyon shook his head. Galen sighed.

"Funny thing. Nobody else seems to, either."

Galen's expression never changed. But the air between them was now frigid. It took Semyon a full second to respond.

"Are we finished here?"

"It's just a matter of time, you know. Before I figure out who you really are. So enjoy yourself, because your days as the Darling of Las Vegas are truly numbered."

He gave Semyon a last look, then stepped back and let the elevator doors slide shut.

Alone in the elevator, Semyon leaned back against the wall. He was breathing hard. Victor had been wrong. It had been idiotic for Semyon and Owen to come back to this casino. Semyon had to find Owen, to tell him that Galen was onto them. It was just too dangerous for them to be there anymore.

He wiped sweat from his brow. Jack Galen wasn't going to give up. He wasn't going to stop until he'd brought the team down. He was making this personal, and that scared Semyon more than anything else. Because when things got personal, emotions got involved. He really didn't want to find out what sort of emotions a seemingly emotionless man like Jack Galen was capable of having.

For the first time since the plane crash, Semyon no longer noticed the pain in his foot.

CHAPTER 25

A Casino on the Strip
Las Vegas, Nevada

en minutes later, Semyon was back in the same elevator, this time heading up, toward the VIP suite. His heart was beating fast in his chest, and he was still breathing hard. He'd spent the past ten minutes frantically searching the pool area for Owen, to no avail. It hadn't helped matters that the roped-off section had been packed with drunken, half-naked partyers—a fitting scene for a Vegas casino setting, but a hellish inconvenience for someone trying to locate a friend in potential danger. Semyon hadn't seen Victor or Allie either; he just hoped that they hadn't headed to the blackjack tables. He was sure Galen was back in some security booth, watching through those fucking cameras. *Always watching.*

The elevator opened on the VIP floor, and Semyon rushed out into the hallway. His cane bounced against the carpet as he made his way from door to door, the numbers blurring together as he searched for Owen's room. The wedding suite had come with four adjoining suites; Owen had claimed one for himself when they'd first arrived, and once Victor had caught sight of Owen's date for the evening, he'd let the claim stick. Semyon figured that if Owen wasn't down by the pool, he had probably stopped off in the room to get better acquainted with his "friend." Semyon didn't relish the idea of barging in on Owen, but this was indeed an emergency. He didn't think Jack Galen would think twice before interrupting.

Semyon rushed to Owen's door and stopped suddenly, using his cane to keep himself from teetering over with the momentum. To his surprise, he noticed that the door was a few inches open; soft jazz was leaking out into the hallway, and there was the distinctive smell of incense in the air. Semyon gritted his teeth, then placed his palm against the partly open door, giving it a soft shove.

The living room was dark, lit only by the glow of the city reflected through the large picture window. Semyon entered slowly, his eyes adjusting to the difference in light. He slowly made out a leather couch, a coffee table, and a large-screen TV. The TV was off, but even if it had been on, the picture would have been obscured by the tuxedo shirt that was hanging over most of the screen. Moving deeper into the room, Semyon could see the rest of the tuxedo draped over the couch, next to the crumpled material of a woman's cocktail dress.

Shit. Well, at least he'd found them. He took a deep breath, searching for the source of the incense. The smell seemed to be coming from the direction of the bathroom.

He set his jaw and moved to the small hallway that led out of the living room and deeper into the suite. As he turned the corner, he saw a soft glow coming from an open door—and heard the sound of splashing. He moved quicker, his cane tapping against the floor.

"Owen?" he said as he entered the bathroom. "Sorry to interrupt—"

And then he stopped in the doorway. The bathroom was spread out in front of him: tiled floors, a double sink, mirrors on the walls. There was a huge bathtub along the far wall, filled to the brim with soapy bubbles. Candles and sticks of incense had been lit and placed on almost every available surface, the glow reflecting off the mirrors, ceiling, and walls. Semyon's attention was immediately drawn to a figure rising out of the bubbles—and it wasn't Owen.

"Hello there," Kimberly said, a wide smile on her puffed-out lips.

She was naked, save for the shine of soap on her overly tan skin. Her arms were crossed against her ample chest, but the gesture did little to cover her curves, instead bolstering the display. Her stomach was flat, a tan glade leading down into the bathtub, where she was on her knees, bending slightly forward, her eyes as inviting as her voice.

Semyon coughed. "I'm looking for Owen."

"He's not here," Kimberly said. "He's across the hall. He left a key on the couch, but I think he's mad at me about something, so I'm staying here."

She was obviously drunk. Semyon took a step back—and she rose higher out of the tub, revealing more of her shining skin.

"Don't go. Why don't you come over here and join me for a while. Owen won't mind. We have an open relationship."

She laughed at this, and Semyon guessed they didn't really have much of a relationship at all, at the moment. He shook his head.

"Thanks for the offer, but I've got to find Owen—"

Before he could finish the sentence, she was out of the tub and coming toward him. Her arms came out and suddenly she had him up against one of the mirrors, her body pressed hard against his chest. He could feel the warm water soaking through his clothes, feel her wet curves rubbing up and down against him, and he tried to push her away.

"Wait a minute," he said feebly.

Then her mouth was on his and her tongue was pushing against his lips. He felt her hand go down the front of his pants, her fingers closing—and then there was a voice, female, from the doorway behind him.

"Semyon?"

Allie. Semyon grabbed Kimberly's arms and pushed her back. He turned and saw Allie standing in the doorway, squinting in the dim light. She was still in that silver dress, a bottle of champagne in her right hand. Her eyes met Semyon's, then shifted to the girl in front of him. Then she shrugged, turned, and walked out of the bathroom.

"Allie, wait," Semyon said, struggling to untangle himself from Kimberly. By the time he had worked free, setting the drunk and naked girl back into the bathtub and made it out into the living room, Allie was gone.

Fuck. That was just what he needed. Allie to see him with Owen's stripper all over him. It was going to be a mess to explain. Semyon cursed to himself as he quickly searched the couch for the other room key. He found it beneath Owen's pants, then quickly hobbled toward the door.

He'd chase down Allie after he'd warned Owen about Galen. Hopefully, she'd understand. He quickly exited the suite, crossed the hall, and placed the key into the door of the room directly adjoining the wedding suite. He didn't know why Owen had left Kimberly alone in the bathtub, or what he was doing running around the hotel without his clothes. He wondered if he was the only one left in the whole casino who was sober.

He opened the door—and found himself face-to-face with a woman in leather pants and a white lace halter top. The woman had bright red hair, tons of makeup, and a chest that rivaled Kimberly's. She smiled at him, then brushed past, heading quickly down the hall. He looked after her, watching the leather pants sway back and forth as she rushed toward the elevator.

What the hell? She definitely hadn't been one of the wedding guests. If Semyon had to guess, she was a hooker. This was Vegas, after all, and you saw them everywhere. In the hotel bars, on the sidewalks outside, even on the casino floors. Certainly, around the high-stakes lounges. The question was, what was she doing in the suite?

Semyon stepped into the room, shutting the door behind him. He didn't have to look far to find Owen; his friend was sprawled out on the leather couch in the middle of the room, stark naked. He was about to turn away, to give his friend time to cover himself with something—when he saw the yellow rubber rope, tightly tied around Owen's bare arm, just above the elbow.

Semyon's eyes went wide, his chest constricting. He took a step forward. Owen's face was toward him, but his eyes were closed. His cheeks looked pale, almost white, and there were beads of sweat on his forehead. Semyon's gaze shifted to the floor—and there it was, the syringe, glistening in the neon light from the picture window.

Semyon rushed forward. He reached the couch and grabbed Owen by the arm. Thankfully, Owen's skin felt warm. Semyon untied the rubber string, tried to ignore the small bead of blood dripping down from the spot where the needle had been inserted. He touched his friend's cheek—and Owen's eyes came half-open, the lids heavy and swollen.

"Hey, man," Owen said. "Hell of a fuckin' party."

And then Semyon's concern melted away, replaced by red hot anger. Owen wasn't in trouble—he was high. Completely fucked up on heroin. Probably given to him—or sold to him, more likely—by the redheaded hooker.

Victor had been right about Owen. Semyon let go of his friend's arm and stepped back from the couch. Inside, he was seething. He had protected Owen, hidden his seventy-five-thousand-dollar mistake. He had ignored Victor's warning, had given Owen the benefit of the doubt. Now Owen was here, drugged out in the casino. If Galen found him like this, he'd have them all arrested—for a real crime that had nothing to do with cards.

Semyon took a step toward the door. The only thing he could do was get the hell out of there. Let Owen sleep it off, pray that Galen never found out about the heroin. There weren't cameras in the hotel room—that would be against federal law—so if Owen stayed in the suite, he'd be out of Galen's reach.

"Man," Owen said, from the couch, "I fucked up again. I really fucked up again."

Semyon paused, lowering his head. He couldn't just walk out on Owen. As much as he wanted to, he couldn't just leave him like this.

"Yeah, you did. But you'll be okay."

"No, man, not this. I fucked up. With the girl."

Semyon rolled his eyes, then moved back toward the couch. He took one of the pillows and placed it over Owen's lower half, covering him up.

"Kimberly will get over it. She seems like the understanding type."

"Not her," Owen mumbled, his eyes drifting close. "The hooker. I fucked up with her. I told her about the money."

Semyon stiffened.

"What money?"

"My stash. For the weekend. I had it with me. In here, behind the

TV. I needed some of it to pay for her, so I brought it in here from the other room and hid the rest behind the TV—"

Semyon was up and moving across the room. He dropped to one knee, shoving at the huge television set with both hands. He searched the space behind the thing with wild eyes—but there was nothing there. He dropped back, still on his knees, and looked at Owen.

"How much money? Owen, how much money did you hide behind the TV?"

Owen didn't answer. He was nodding off. Semyon rose to his feet.

"Owen, how much goddamn money!"

"Fifty," Owen mumbled. "Fifty thousand."

And then he slumped back onto the couch, passed out.

Semyon collapsed back against the television set, staring at his drugged-out friend. Owen had hired a hooker, bought heroin from her, shot up, then let her walk out the door with fifty thousand dollars of the team's money.

Semyon shut his eyes. Maybe Jack Galen wasn't the worst of their problems.

CHAPTER 26

Hard Rock Hotel
Las Vegas, Nevada

PRESENT DAY

She was tall and blond and beautiful.

She was young, maybe nineteen or twenty, with smooth, pale skin, long limbs, and pretty green eyes.

Her hair was long and straight, hanging down to her shoulders and across her forehead in gently teased bangs. She was wearing too much makeup, but the stark contrast of the bright red lipstick and ocean blue eye shadow against her innocent features was somehow endearing, a girl trying to be a woman but landing somewhere in between. She was smiling, but she was also a little nervous; her hands were trembling and her small, natural chest was rising and falling beneath her slinky black dress. She was looking in my direction, but not really at me, more at her own reflection in the window behind me.

Her name was Skylar, and she had just asked me if I wanted her to take off her clothes.

Ten seconds later, the question still hung in the air of my hotel suite, melding with the thick, flowery scent of her perfume. It was the sort of question you expected to hear in a suite like this; the place was decked out like an eighties rock-and-roll crash pad, with leather furniture, thick carpeting, dim mood lighting, and a wraparound picture window glowing with the reflected flashing neon image of a massive purple-and-pink guitar. It was one of the Hard Rock Hotel's best rooms, and it was designed for moments like these.

Except, this time, the moment wasn't exactly what it seemed.

"Actually," I finally responded, sitting on the edge of one of the leather couches, "I just want to talk."

God, it was such a clichéd thing to say, and I was certain she'd heard it many times before. Maybe not a little after midnight on a Friday night, in a top-floor suite in the Hard Rock Hotel, but enough times for her to realize that it was usually a less than sincere way of softening the wrongness of the moment, of keeping some semblance of control over an inherently out-of-control situation.

"Okay," she said, crossing to a matching couch set directly across from where I was sitting. "Sure, whatever you want. But it's still going to be three hundred dollars."

I reached into my pocket and withdrew the bills. Hundreds, three of them, because this was Vegas after all, and I'd spent enough time with Semyon to have had some of his habits rub off on me. Even to this day, he still carried hundreds, and especially in Vegas. It was a character trait that linked all high rollers, and tonight, I was living the high-roller lifestyle, at least in appearances. Semyon had made sure of that.

The suite was comped to one of the few aliases that still worked for him, provided he never actually showed his face in the casino and allowed the oblivious host a chance to match a face with the name and betting record of the high roller who had earned such VIP treatment. The flight from New York had also been gratis, donated by a completely different casino that expected me to meet another one of Semyon's still-viable aliases for a round of high-stakes blackjack at some point during the weekend. The wildly expensive meal I'd had at Nobu downstairs in the casino had been picked up as well, though I wasn't certain which alias had done me the favor, and could only assume it was the same version of Semyon who was picking up the lavish suite.

Still, the suite, the meal, and the flight were all just a prelude to Semyon's introduction to high-roller living; he'd wanted me to enjoy the real experience, down to the details—which is why I hadn't been surprised when he'd called me ten minutes before Skylar had arrived, informing me that I'd be entertaining a guest for the next hour or so. Unlike the suite, the flight, and the dinner, Skylar's services weren't gratis; but the appointment itself had been arranged with the same ease, through the same sources, as the rest of the high-roller comps. Sure, the casinos do not endorse the procurement of hookers, but they understood, better than anyone, that Vegas was supposed to be a full-service experience. And to keep a high roller's business, an overeager host might bend the rules.

I looked at Skylar sitting across from me, her long, bare legs crossed at the knees, her high heels bouncing up and down as she leaned back against the leather couch. I wondered how many hotel rooms she was going to visit tonight, how many rooms she'd already been to, how much money she'd make this Friday night. Three hundred dollars didn't sound like much, but it was only the amount that got her to the door. Extras were extra, and I knew that for most of the men who called for her services, it was all about the extras. If you just wanted to see a naked girl, there were a dozen strip clubs within ten minutes of this hotel where you could see hundreds, if not thousands, of naked girls. There was a reason you wanted them naked in your room.

"So what do you want to talk about?" she said, after I placed the three hundred dollars on the coffee table between us. I knew she was expecting that pile of hundred-dollar bills to grow over the next hour, but she had to work her way through the conversation first. Certainly, she'd been through this many times before.

The truth was, you didn't have to be a high roller to get a woman like Skylar sent to your hotel room in Las Vegas. The truth was, it was much easier to get a beautiful girl sent to your room in Vegas than a quality pizza—and the girl would arrive in half the time. One phone call, fifteen minutes—and Skylar was sitting across from you, waiting for you to ask her to take her clothes off.

Nobody had an exact figure on the number of sex workers in Las Vegas, but estimates ran into the tens of thousands. If you included every level of the sex industry—from the strip clubs, to the street-walkers, to the room-service escorts like Skylar, the numbers might have run even higher. Without question, it was a massive, nine-figure industry, one that was growing astronomically every year. Some experts had prophesied that it would be in the billions within the next few years.

"I'm just curious about what you do," I said, picking my words carefully.

"You're not a cop, are you?" she asked. I smiled, shook my head.

"A writer, actually. I'm working on a story."

She smiled back. I'd found that in some instances, the truth actually worked, and I wasn't averse to using it when it would play to my advantage. Girls like Skylar weren't afraid of journalists because often in their minds, what they did for a living was a script they were living out, a novel—comic, tragic, romantic—that they were writing with their lives. They didn't fear journalists, but they did have reason to fear cops.

Despite what many tourists believed from the plethora of billboards and leaflets promising naked girls in your room, anytime, anywhere, prostitution was, in fact, illegal in Las Vegas. The legal brothels, where prostitution was officially sanctioned, were located an hour outside the city, across Clark County lines. However, although prostitution was illegal in the city itself, paying a girl to come to your room and take off her clothes was not considered prostitution. A loophole that became a doorway, because once a girl was naked in your room, what happened next was fairly elementary.

"Well, there's definitely a story to tell," Skylar said, running a hand against her bare knee. "Poor little small-town girl caught up in a dirty world, that the kind of thing you're looking for?"

I shrugged. I had no doubt that it was a dirty world, but I doubted that Skylar was the babe in the woods that her innocent features implied.

"Something like that."

She laughed. "Well, it wasn't a small town, it was more like a suburb. I was working as a bartender outside of Detroit, trying to put together enough money to pay for school. Well, every night I got hit on by guys from the minute we opened to the minute we closed. Then one weekend I came to Vegas, and one of my male friends took me to a strip club. I watched those same guys who hit on me putting twenty-dollar bills in the girls' G-strings. So the next weekend, I flew out to Vegas on my own and started stripping. First the Paradise, then the Crazy Horse. I was making good money, but not enough for school. Then I met a girl who introduced me to a friend of hers who ran an escort service. He worked for another guy who ran an even bigger escort service. They told me how much money I could make, and I didn't believe them."

The slippery slope from stripping to hooking. I had a hunch that there were a lot of girls like Skylar who had made the jump from dancing naked to escorting, though most strippers you talked to said there was a huge difference between the two industries. At least, that's what they liked to believe.

"So how much money can you make?" I asked. I didn't expect the truth, but I was hoping for something close.

"In a night? Ten thousand dollars. Sometimes more."

She was probably exaggerating, but even if she'd doubled it in her head, it was still an enormous amount of money. I didn't want to try to calculate how many guys she had to sleep with to make that kind of money.

"How many girls work for your service?" I asked.

"Maybe fifty."

Fifty girls making five to ten thousand dollars a night—that would add up to something between a quarter and half a million dollars being generated on a Friday night for the escort service that employed them. To put it in perspective, the MGM Grand, a billion-dollar hotel, made about three million dollars of profit on a Friday evening. And the MGM was regulated by a gaming commission and answered to a sea of stockholders.

"Is it dangerous work?"

"Not if you're honest. If you're prone to stealing, it can be. The people who run the service can be kind of nasty."

I realized she wasn't talking about stealing from the johns. I'd done a fair amount of research into the escort industry, and I'd come to a conclusion: The people who say there's no mob in Las Vegas anymore are dead wrong. The mob is still around; they're just not involved with the casinos. They were kicked out of the biggest industry in town, and instead have taken hold of the second biggest industry: sex. And that was the ultimate growth industry.

The truth was, sex was rapidly overtaking gambling as the main reason people came to Las Vegas. Even the Las Vegas Board of Tourism had tacitly admitted the fact, with its wildly successful recent advertising campaign, centered on the catchphrase *What happens here stays here.*

Did anybody really think the ads were talking about the five hundred dollars you lost at the blackjack table? Or the five hundred dollars you spent to get a nineteen-year-old girl sent to your room?

Obviously, the Las Vegas Board of Tourism had finally asked itself: Why was Las Vegas the premier bachelor-party destination in the world? Because of the gambling? And the answer was easy: because of the strip clubs and the escort services.

"So you decided to stick with this," I commented. "Make some serious money. You moved out here to Vegas—"

"No, I still live outside of Detroit. I fly out here every Thursday and back every Sunday. I'm taking classes during the week in restaurant management, and I still live at home with my parents. They think I have a boyfriend in Los Angeles who's paying for my school."

I raised my eyebrows.

"So you commute to Las Vegas to work."

She shifted against the couch, uncrossing her legs, then crossing them again. I tried not to let my eyes follow the motion too closely.

"Most of us do," she responded, playing with a loose strand of blond hair that was hanging down over her bare right shoulder. "The escorts and strippers. You ever look around McCarren Airport on Sunday night? You ever notice that there are a whole lot of girls in sweatpants with boob jobs and bad hair waiting for flights?"

In my head, I saw all those airplanes full of nubile young women flying into Vegas from cities all over the country. I saw the thousands of beautiful girls being guided by air-traffic control toward the neon glow in the desert. I realized that at that very moment, there were young women getting naked all over Las Vegas. It was a numbing thought, but the truth was, many of the men who came to Vegas thought of prostitution as part of the Vegas experience.

What happened to Owen in that hotel room after the wedding was unique only in that he'd been stupid enough to brag about his blackjack stash and had gotten himself robbed of more than fifty thousand dollars of the team's money. But partying with an escort was as normal in this town as ordering room service. Many visitors to Las Vegas had taken the tourism board's slogan to heart; they believed that they were supposed to leave their inhibitions at home, they were supposed to give in to the sins that made Vegas seem so delectable. And that was exactly what Las Vegas wanted them to believe. The two most basic sins of human nature were gambling and sex—the linchpins of the neon economy.

I ran my hand over my eyes, thinking about everything she had told me. Trying to sync it all together with the image of the pretty young blond girl sitting across from me, with those innocent features painted over in the lush war paint of her profession. Then I noticed that she was looking at me, and her expression had changed—a little less innocent, a little more Vegas.

"Well?" she asked, showing the tip of her tongue between her overly red lips.

"Well what?" I asked back, confused.

"Now that you've got your story, do you want me to take off my clothes, or don't you?"

I stared at her, playing with that strand of blond hair, her bare legs crossing and uncrossing against the leather couch.

There was a short answer to her question, and a long one. But the best answer was the one that had been graciously provided to me by the Las Vegas Board of Tourism.

What happens in Vegas stays in Vegas.

CHAPTER 27

Boston, Massachusetts

he first thought that hit Semyon as he carefully navigated the descending wooden stairs that led to the basement apartment in Somerville was that Victor should have sold a few of his suits and found someplace nicer to live. It wasn't that the apartment was awful; from the third step, Semyon had a good view of the place, and it seemed habitable. Wall-to-wall carpeting, built-in shelving units covering most of the walls, an alcove with a futon in one corner, a galley kitchen in another—pretty much what you would expect from this blue-collar city adjacent to Cambridge and about twenty minutes from the MIT campus. But "habitable" wasn't what you expected from the guru of a multimillion-dollar blackjack team, even if he was notoriously cheap.

Then again, Victor had crashed an airplane into a brick wall because he was too stingy to take a full course of flying lessons. Semyon realized he should not have been surprised by a basement apartment.

By the time he'd reached the bottom step, he could see that the rest of the team was already there, gathered around the living-room table—which was actually a glass door set on its side, atop a small packing crate. On top of the table, as a centerpiece, Victor had placed the same metallic contraption Semyon had seen during his introduction to the team—the fake-ID machine. There were a pile of new IDs next to the machine—but after the face-to-face incident with Jack

Galen, Semyon knew it would take more than new IDs to make him feel comfortable anywhere near Las Vegas again.

As he approached the couches, his gaze drifting from Victor and Jake on one side of the coffee table, to Owen, Allie, Misty, and Carla on the other, he went over the events of the weekend before in his head, the images flying by at warp speed. Jack Galen, his thin fingers holding the elevator door open. The naked girl coming out of the bathtub, then minutes later Allie catching them in a sitcom-worthy moment. Then Owen, drugged out on the couch, telling him that he'd just been rolled by a hooker who'd walked off with fifty thousand dollars of the team's money.

He wasn't even sure how to prioritize the mess in his head. Allie seemed the simplest and most absurd tragedy of the weekend, and he'd already spent most of the flight back to Boston trying to fix that one up. He wasn't sure that she'd believed his account of the bathroom scene, and she'd acted as though she didn't really care—but he had a sinking feeling that the episode had set back his chances of reaching her in a more significant way anytime soon.

As for Owen, Semyon had pretty much avoided his friend altogether since they'd left Vegas. He hadn't said anything to Victor about the stolen money, or the hooker, or the drugs—but he wasn't exactly protecting Owen anymore, he just wasn't going to be the one to bust him. He wished it could be different. But he'd seen the evidence with his own eyes. Owen was a loose cannon.

And even without Owen, Semyon had the feeling that in Vegas, they were living on borrowed time. He had told Victor and the rest about Jack Galen; Victor had pretty much blown the incident off, maintaining that the whole idea behind the three techniques was that you had to come in strong, make a scene, control the blackjack table as well as the casino itself. The house was supposed to know you, to wonder about you, to keep your face and your play style on file. You weren't supposed to be flying under the radar, you were supposed to be blowing right through it. So of course you were going to get heat.

But Semyon knew that Galen was something different. He wasn't just some suspicious pit boss, or even a hungry private eye. He was a man on a mission. He was dangerous.

Still, Victor's logic had, for the time being, prevailed. After the dust settled after the blowout wedding, the team had played the rest of the weekend. Mainly at the MGM Grand and at the Stardust, but also at the Flamingo and Circus Circus, a more family-oriented establishment that wasn't as used to high rollers but was willing to take the action. They'd run ace sequences at the MGM and had won two hundred and thirty thousand dollars. At the Stardust, they'd busted the dealer twelve times with the third technique and search-cut to over twenty aces using the first technique, and had won a total of four hundred and thirty thousand dollars in two days. At the Flamingo, they'd sequenced and cut to another hundred and thirty thousand, and at Circus Circus, they'd really broken the bank, winning nearly half a million dollars in eight hours. All told, the team was up close to two million now, in a little over five months. It was an incredible run, and there was no doubt that these techniques were more powerful than anything that had ever been seen by the casinos before.

Which was the real reason Victor had called this meeting in his basement apartment and had the ID-making machine pumping out new IDs at full tilt. The time was ripe, he'd explained, for them to try something big. No, not just big—something spectacular.

He wanted to take the team back on the road. But not to Atlantic City or the Caribbean. Victor wanted to do something memorable, something that would embed them, forever, in the annals of gambling lore.

"Europe," Victor said as Semyon sat next to Allie on the edge of the couch, his good foot braced against the glass tabletop. He didn't need his cane anymore, but his right foot was still weak enough to cause him to favor his left. "Start in Barcelona, then move on to Amsterdam, then to London—and it all leads up to the ultimate target."

He held his hands out like he was offering them something precious and important.

"We're going to hit the most famous casino in the world. Monte Carlo."

The room didn't go exactly silent, but pretty close. Owen was inspecting his leather jacket and Allie was playing with her hair. Carla was twisting her wedding ring back and forth and Misty was adjusting a bra strap that had peeked out from beneath her open-necked sweater. Jake was just breathing, but that was actually louder than anything else in the room, the pre-asthmatic huff of an obese young man. But it was up to Semyon to truly break the quiet, and the tension, by saying what they were all thinking.

"Victor, that sounds like fun, but it also sounds insane. I mean, I agree that we should take some time off from Vegas—we've made a lot of money, and we're too hot for most of the casinos in town. But we're going to be real fish out of water in Europe. I mean, look at us. Our acts work in Vegas and AC because those places are all about the show. They're expecting the scenes that we create. But from what I've heard about Europe, and especially Monte Carlo—it's just different."

Semyon was far from an expert, but he'd read about Monte Carlo during his training period, and he knew that the famous old Monte Carlo Casino—the oldest casino in the world—was nothing like the plastic fantasies of Vegas. This was the casino from James Bond movies, where European royalty in tuxedos and ball gowns congregated to sip champagne and bet millions on games like baccarat. Would a bunch of MIT kids be able to pull off their act in a place like that? In a country that was ruled by a monarch and thought of Americans as ugly, pesty insects wearing shorts and sneakers, haggling over the price of a baguette?

Semyon wasn't an expert, but he also hadn't been born in the United States. He knew that the rest of the world was very different. As bad as Vegas and AC could be to card counters and suspected cheats, he knew that this country was paradise compared with other places around the world. It wouldn't just be hard to pull off—it would be dangerous.

"Okay," Victor said, nodding. "It's definitely going to be trickier. In some of those casinos, we'll need to show passports, and we can't fake those. And we'll have to go in and out much quicker, maybe make a little less noise. But I know we can pull it off. And come on, deep down, don't you want to try and do something great? Something really . . . historical?"

Semyon had to admit, Victor was a compelling speaker. There was something to what he was saying—about making history. They'd already made an enormous amount of money—and as an equal partner, Semyon was going to get rich if things continued. Hell, he was rich just on what they'd made so far. But hitting Monte Carlo would be something else. But was it too dangerous?

Jake was the next one to speak up, and Semyon could tell from his voice— and the fact that he was looking across the table at Carla, not at Victor— that he'd already made up his mind.

"Sorry, man, but Carla and I are out. And Misty, too. Carla and I have decided to take time off, now that we're married. Not just to recover from the wedding, but to find an apartment, get our shit together, adult stuff. And Misty's going to help us with the move. But we expect a lot of postcards."

He grinned, and Victor nodded, smiling back. He wasn't going to force anyone else to join his party. Semyon had a feeling that he'd go on his own, if he had to. Or recruit a new team for the endeavor. He still had no idea what drove the man, but Victor was more than just a control freak, and it was more than just about money—like Galen, Victor was a man on a mission. Semyon just hadn't figured out yet what that mission really was.

To his surprise, the next person to speak up was Owen.

"I'm in. I've always wanted to check out Amsterdam. And I'd love to give those Euro shits a taste of our good old American ingenuity."

Semyon caught Owen's eye for a brief second, then looked away. Of course Owen was in. He had a debt to repay the team; he *had* to keep playing. Maybe he was hoping they'd hit it so big in Europe, Victor would never notice that there was fifty thousand unaccounted for from the last trip to Vegas. Or maybe Victor already knew about the money, and Owen was trying to stay in his good graces by backing him on his idea. But Owen's response obviously had some effect on Allie because she suddenly raised her thumb and winked at Victor.

"I can't think of a better way to see Europe than as part of a blackjack invasion. Who knows, maybe some European prince will come and sweep me off my feet."

Semyon wasn't sure, but he thought Allie glanced at him as she added the bit about the prince. He wondered if the flippant comment had been aimed at him. He grimaced inwardly, then sighed out loud.

"Okay, Victor. You can count me in, too. But we're going to be a lot more careful than we were in Vegas or AC. I don't think the golf courses in Monte Carlo are quite as hospitable as the ones in Aruba."

Victor grinned at him, the decision made. In that moment, Semyon could tell from his eyes that Victor had never even considered that his team would turn him down. Semyon was bothered by the thought.

The truth was, even though they were partners, they both knew who was still running the show.

Outside, a cold breeze whipped across the sidewalk, whistling between the matching row houses that lined both sides of the quiet neighborhood street. Semyon was standing on the sidewalk, pulling the collar of his winter coat tight against his throat, trying to stay warm. Owen was next to him, but he didn't seem to care about the cold. His leather jacket was open, his hands hanging limp at his sides, and he was obviously trying to come up with the right set of words to make things okay again.

Semyon hadn't intended to give Owen the chance, not now, not yet. He'd actually been trying to catch up to Allie after the meeting had concluded, but by the time he'd gone over some of the basic planning of the European trip with Victor and left the basement apartment, Allie was already firmly planted in the backseat of Jake's car. Jake and the three girls had since taken off, leaving Semyon and Owen alone on the sidewalk, waiting for the cab that Victor had called for them.

Finally, Owen built up the courage to say what needed to be said.

"Okay, Semyon, I know I fucked up. And I'm sorry. It's not like I use the shit all the time. The girl had it on her, I was drunk, I did something stupid. I haven't touched the stuff since."

Semyon rubbed a free hand over his eyes. Owen's voice sounded sincere, but he didn't know if he could believe him. Two times, Owen had screwed up, losing the team's money. Maybe drugs had been involved the first time, too. Certainly, he'd been arrested before because of them. Maybe he was an addict, couldn't control himself. Maybe Semyon was doing the wrong thing by protecting him, by helping cover it up.

"I'm not going to tell Victor," he responded. "But I'm not going to lie for you either—"

"I'm not talking about Victor," Owen said. "I don't care about Victor. I'd fucking walk away from this team in two seconds, and never look back. But you're my friend. I want you to know that I'm sorry, and that I won't do it again."

Semyon let go of his collar, letting the cold wind touch his throat. He could feel the emotion in the air. Owen was being honest. He was putting it out in the open—he valued Semyon's friendship more than he valued the team. Semyon knew it was probably a mistake— but he wanted to trust Owen again. He wanted to forget about what he had seen in that hotel room.

"Owen, you've got to be smart from here on out. I don't care what you did in the past—but if you blow it again, I will kick you off this team myself."

Owen nodded, and then Semyon held out his hand. They shook, and Semyon allowed himself to crack a smile.

"And the next time you leave a naked girl in a bathtub for me, at least have the courtesy to give me a heads-up."

Owen smiled back, and their handshake turned into a hug. Semyon hoped he was doing the right thing. Another decision made, the second important one of the day. Forgive, try to forget, because the team was taking their game to Europe. Heading toward the ultimate conquest—the most famous casino in the world.

CHAPTER 28

The Gran Casino, Barcelona, Spain
The Holland Casino, Amsterdam
The Ritz Club, London, England

urope in flashes, like postcards caught in a vicious wind, everything moving so fast that every time Semyon blinked, the images changed. Swirling visions flashing, a blend of vivid, if anachronistic, pictures that stood out amid the fog of jet lag, the blur of cards spinning against a vast sea of green blackjack felt.

Barcelona: a place of mind-numbing contrasts, of Gothic spires and modern curves, of ancient stone and tempered glass. Semyon, Owen, Victor, and Allie navigating the narrow streets of the ancient Barrio Gothic, a medieval spiderweb of alleys originally built by the Romans and still seemingly untouched by time. Making their way back and forth between the short-term rental apartment Victor had found through a friend of his back in the States—a place so down-trodden and disgusting that Semyon knew it had to be the cheapest place in the entire city, its only real selling point the in-room safes where they could stash their loot—and the Gran Casino at the Port Olympic, a smoke-filled, old-world casino lodged in a strange architectural complex beneath a more modern skyscraper that over-looked the Mediterranean. Afternoons spent touring the walkable open markets, sitting at waterfront cafés sharing tapas and drinking sangria, avoiding the pickpockets in the tourist traps of the old city, picking their way through the Museu Picasso and the unfinished La Sagrada Familia, a church with lavish bell towers covered in Venetian mosaics. All of them feeling out of place, every minute of every day, and yet loving it, because unlike the thousands of tourists they passed in the streets and in the churches and museums, they were in Barcelona for a reason: night, when, like predators, they roamed the dark pits of the casino, plying their trade.

The Gran Casino's focus was on the European table games; roulette, baccarat, mini–*punto banco*, craps. The first night, it took Semyon and his team ten minutes to find the five high-stakes twenty-one tables, lodged as they were in a dark alcove near the back of the sprawling building. But the table limits were high enough, and more important, the dealers were completely oblivious; all of them men, Spanish-speaking, olive-skinned, smiling, and small of stature. During one hour of play, sitting in the first-base seat, Semyon had seen the back card of every single shuffle and had caught three aces, all of which he'd steered to his teammates' hands. He'd busted the dealer twice with a king and a jack. Though the betting limits worked out to less than a thousand American dollars a hand, they had easily gone up nearly thirty grand by the end of their first night. The only real problem they'd run into was the notion that they show a passport at the casino cage to exchange their chips. Since at least one of them would be playing under a real name, they had to play more carefully than in Vegas; keeping their bets down when the pit bosses were near, keeping the theatrics to a minimum. But even so, catching a few aces an hour and busting the dealer when possible, they would still be able to take the Gran Casino by storm.

However, on their second night in Barcelona, Semyon was able to solve the passport problem—quite by accident.

It was 2:00 A.M., and the jet lag had Semyon staring at the cracked ceiling of the crappy apartment, listening to Owen snoring in the twin bed on the other side of the small, cell-like room. As usual, he was thinking about Allie; there hadn't been any chance for one-on-one time since they'd left Boston, and all he really wanted to do was go across the hall to her room, take her by the hand, and share a stroll along the romantic streets of the old-city district. Tell her that he thought about her all the time, that since the night of the wedding she'd dominated his thoughts even more than blackjack.

He actually made it all the way to the door of her room, standing there like a fool in the narrow hallway, his hands jammed into the pockets of his bulky jacket, staring down at his shoes. He just didn't have the guts to wake her up and tell her how he felt. Crazy, that he could stare down a pit boss, throw his feet up on a blackjack felt and shout "Pay the table," but he couldn't tell a girl that he liked her. Christ, he was MIT to the core.

Instead, he made his way down the hall to the creaky little elevator that led down to the street. Maybe a stroll, on his own, through the Barrio Gothic would give him the inner strength to say what he needed to say.

The elevator button didn't light up when he hit it, but he could hear the gears and cables groaning on the other side of the accordion-style caged door, indicating that the thing was still somehow working. He contemplated taking the stairs instead, but he'd seen the junkies congregating in the open first-floor doorway on his way into the building, and he didn't think the stairs were going to be any safer than the steel coffin of an elevator.

There was a last metallic groan, and the elevator arrived, coming to a stop behind the accordion cage. The inner metal doors slid open, and to Semyon's surprise, there was someone standing in the back corner of the little coffin, slumped slightly forward. Semyon thought he recognized the young man from the gang of junkies who hung out downstairs: dirty blond hair, rail-thin body, with nervous eyes and scruff covering his triangular chin. But he couldn't be sure.

He thought about turning around and heading back to his room—then decided that the scraggly guy didn't look too dangerous. He was practically swimming in his baggy jeans and dirt-stained sweatshirt, and Semyon thought if any trouble broke out, he'd probably be able to hold his own.

He pulled back the metal accordion and stepped inside. The entire elevator reeked of alcohol, and he had to fight to keep from gagging at the heavy scent. He tried to stand as far away from the kid as possible, but the minute the elevator started to descend, he realized that the kid was inching toward him.

"Hey, man," the kid said, in English but with a heavy Spanish accent. "You got a cigarette?"

Semyon shook his head. He could see that the kid was shaking, way more scared than he was. That could be a good sign, or a bad sign.

"How about money? So I get something to eat?"

The kid was staring at him now, with a hungry look in his dark eyes. But Semyon didn't think he was hungry for something to eat. Semyon checked his pockets, found a dollar bill, and held it out for the kid.

"This is all I got." Actually, he had a bit more than that on him. There was a thousand dollars in his back pocket, in a roll of ten hundreds, but he had no intention of handing that over. Not without a fight.

The kid grabbed the dollar bill, then glared at Semyon.

"Come on, man. I get shit for that."

The kid's hand came out from under his sweatshirt. Semyon saw a flash of metal and realized the kid was holding a knife. Not a big one, really little more than a pocketknife—but it was menacing just the same. Semyon held out his hands, backing against the wall.

"Whoa, let's not get crazy. You robbing me?"

The kid's face was halfway between menacing and apologetic.

"Sorry, man. I need more than dollar. You have more than dollar!"

Semyon thought for a brief second. Was it worth getting cut up to stand up to this jackass? He came to a quick decision, reached into his back pocket, and pulled one of the hundred-dollar bills loose. Then he held it out in front of him. The kid's eyes went wide, and a smile broke out on his lips.

He grabbed the hundred, and for a brief moment Semyon actually thought he was going to embrace him in a massive hug.

"Man, thank you . . . I sorry, but thank you."

The elevator came to an abrupt stop on the first floor, interrupting the kid mid-genuflect. He turned and immediately went to work on the accordion door, turning his back on Semyon. For a brief second, Semyon thought about tackling the weak-looking guy from behind, taking his hundred bucks back—but then he saw something that made him change his mind.

Sticking halfway out of the kid's back pocket was a passport. From the color, Semyon could tell it wasn't a U.S. passport, probably Spanish or Portuguese; he doubted the kid was local, probably one of the many kids that age who traveled around Europe looking for highs.

Semyon took a deep breath and waited for the exact right moment. Just as the kid yanked the accordion doors open and stepped forward, Semyon's hand shot out and grabbed the passport by one edge. With an agility born of hours spent practicing with cards, he quickly yanked the passport free and jammed it into his own back pocket. The kid didn't seem to notice. He was too busy rushing out of the elevator, the precious hundred-dollar bill crumpled tightly in his hand.

As he raced down the hall that led to the front door, the kid looked over his shoulder at Semyon, still in the elevator.

"Thanks again, man," he shouted back. "Sorry about the knife."

"Hey, don't mention it," Semyon answered back. And he smiled.

When the kid was gone, he pulled the passport out of his pocket and opened it to the picture. Emilio Díaz stared up at him. With a little facial hair, and a bit of peroxide, it would work just fine. Semyon's smile doubled in size.

A hundred bucks was a pretty cheap price to pay to solve their passport problems.

"This is the life, ain't it, Mr. Díaz?"

Semyon laughed, tossing an empty Amstel can at Owen. They were both lying flat on their backs on the oversize skiff as it bounced along the meandering canal, cutting through the gray morning mist. They'd been out on the water since dawn, because for them God only knew what time it really was; after six days of all-night gambling in Barcelona, a short midnight flight to Schippol Airport in Amsterdam, and three more hours positively killing the Holland Casino, their bodies were now in a time zone all their own.

From his vantage point, Semyon could see the high cobblestone walls that lined either side of the canal flashing by, and above them the pretty Dutch row houses, slammed so close together that the streets of the city were more like rabbit warrens than the roadway of one of the most permissive cities on the planet. Somewhere up there, over the cobblestones, was the red-light district, which they'd strolled through after their gambling route, admiring the half-naked women in the windows, passing the little "coffee shops" where marijuana was legally sold and smoked, wondering why they didn't all just move there and be done with it. Even Allie had joked that MIT should have built a satellite campus in Amsterdam, just to infuse a little life into the student body.

Now that they had taken a moment to relax, as a team, things could not have been better. Long forgotten were the troubles of the past—the barrings in Vegas, the banning from Aruba, the incident with Jack Galen at the wedding. After Barcelona and their gambling session so far in Amsterdam, they had proved that their system worked anywhere, that a group of kids from MIT could take on two of the most famous casinos in Europe and come out ahead—way ahead.

Semyon rose to his elbows. He could just make out Victor at the helm of the boat, hands on the oversize steering wheel. Victor's suit jacket was off, draped over the seat behind him, and his sleeves were rolled all the way up. Semyon was glad that he had turned out to be a much better captain than a pilot; he'd even skillfully avoided a near collision with one of the touristy *bateaux-mouches* that prowled the waterways, without upsetting any of their beers.

"Yeah, this ain't bad," Semyon responded to Owen. Almost all of his concerns about his friend had dried up in the past few days; Owen hadn't even mentioned that they were in the one city in Europe where the use of drugs was not only legal, but in some places, encouraged—and he'd been beyond scrupulous when it came to handling his money. He really had turned over a new leaf, and Semyon felt he could trust him again. Owen still hadn't done anything to pay back the team the money he'd lost—but the way things were going, it wouldn't matter.

Semyon didn't know for sure—Victor had been keeping the written records and would report when the trip was over—but he thought they were now up over two and a half million as a team. And they were going to be hitting Amsterdam for another four days before heading to London. And after London—the ultimate mark: Monte Carlo. In truth, the rest of the European trip was really just a dress rehearsal for the big score. Monte Carlo was the place steeped in legend, the goal of their truly historical mission. And judging from how things had been going, Monte Carlo would fall just as easily as Barcelona and Amsterdam.

Semyon laid his shoulders back down against the skiff, smiling up toward the misty gray sky.

Three days later, two hours after landing in Heathrow Airport— and now definitely *not* smiling—Semyon found himself standing stark naked in a cement walled, windowless room, hands straight out at his sides, while a man with a miniature flashlight and bright green rubber gloves gave new meaning to the term *thorough*. As in, a thorough search by an international customs agent of a suspicious traveler.

The funny thing was—if anything, at that particular moment, could be thought of as funny—was that it had nothing to do with blackjack. And it had only tangentially to do with the three hundred thousand dollars in three different currencies that Semyon had declared to the customs agent as he'd first entered UK territory. In actuality, it had only to do with a stupid little paperback book that Semyon had inadvertently packed in his duffel bag before leaving for the airport in Amsterdam. And in point of fact, it wasn't even Semyon's book; Owen had bought it as a joke, and Semyon had mistakenly packed it in his haste to make their flight.

The book was now sitting on a steel table a few feet from where Semyon was standing. Semyon could see the dog-eared cover from his vantage point, and all he really wanted to do was grab the damn thing and shove it right up the custom agent's—

"Okay," the agent said, rising from behind Semyon while turning off his flashlight. "You've checked out."

Semyon nearly snarled at him as he lowered his arms.

"And I can put my clothes back on?"

"Of course," the man said, smiling kindly. As he crossed the room, he picked up the paperback book and placed it under his arm. "But, Mr. Dukach, we're going to have to confiscate this. And next time you visit the UK, I suggest you leave your pleasure reading behind in Amsterdam."

He shut the door behind him, leaving Semyon alone and humiliated, rapidly putting on his clothes. He couldn't wait to get back to the rest of the team. He was going to kick the shit out of Owen—but even as he angrily yanked his jeans over his legs, he felt a smile pulling at the corners of his lips. It really was absurd. He had three hundred thousand dollars on him, was traveling with three colleagues who had another million and a half in various currencies— and the thing that got him strip-searched was a fucking guide to nude beaches. For some reason, the customs agents had decided that it was some sort of pornographic text—and because it was being carried in the luggage of an American kid with a large amount of cash on him, coming from Amsterdam, they'd assumed that he was some sort of pornographer. It had been enough cause to warrant a strip search—and an embarrassing delay to their assault on London's casinos.

Still, standing there, naked from the waist up, Semyon couldn't help but laugh. They were in the midst of one of the biggest blackjack schemes in history, and the thing he gets busted for is a goddamn pornographic paperback book.

Semyon was still laughing about the episode with Victor, Owen, and Allie as they entered the lavish Ritz Club on fashionable St. James Street, just blocks from Harrods department store and smack in the middle of the priciest district in the city of London. The Ritz, one of Europe's oldest and most exclusive gambling parlors, was a legendary stomping grounds for London's famous and wealthy. You had to be a member to pass through the imposing wooden doorway and into the marble-lined entranceway; thankfully, Victor had managed to convince one of his hosts in Atlantic City to lend him the use of another high roller's name for the excursion. They dressed the part, as well, with suits and ties and Allie in a stunning dress.

Semyon could not help but be impressed as they strolled deeper into the massive mansion. Red curtains swept down from the high ceilings, and the marble floor was polished so brightly that the huge central room could have doubled as a skating rink. Like Barcelona and Amsterdam, the gaming centered on the Euro varieties; the blackjack tables—or American twenty-one, as they were less than affectionately known—were set off from the rest of the casino in a small parlor behind the bar. There were six tables in a small semicircle, set around a raised banker's area. When they arrived, at around four in the afternoon, the room was completely empty. Possibly because it was pouring rain outside, or possibly because it was still early on a Thursday, the entire club seemed at minimal capacity. In the blackjack area, only one dealer was working, a gray-haired man with hunched shoulders and a slender white mustache. He had a bored look on his face as Semyon and the rest approached, and Semyon had the feeling he'd been working the tables at the Ritz for a very long time. Long enough to have seen just about everything. Well, maybe not *everything*.

Semyon chose the first-base seat, letting the rest of the team spread out to his right. Ever since Barcelona, he had been running the actual gaming sessions, if not the European tour itself. After Semyon had showed up in the downtrodden rental apartment with the fake passport, Victor had pretty much accepted the fact that Semyon could run the team as well as he did. And his skills at the table had never been in question. In fact, Victor had even admitted, after one of Semyon's more impressive displays at the Gran Casino, that Semyon had probably become even more skilled at manipulating the bust card than he himself. Semyon had fought back the urge to respond that he'd had plenty of time to practice, lying in bed with a broken foot and burns on the back of his hands.

Once the team was seated at the table, Semyon began pulling the bills out of his jacket pockets. There was no need for any theatrics; as a group of Americans in a private gambling club in London, they were already characters to be noticed, and probably every camera in the place was already watching their game. Especially as the bills piled up on the table. They had exchanged four hundred and fifty thousand of their profits into pounds at the airport—and the way things had been going so far, they didn't expect to need any more to bring the Ritz to its knees. The three techniques, though opportunistic, had proven to work fast—much faster than any card counting scheme. Like spinning straw into gold—or dimes into dollars, as Owen had once put it.

Once their bets had been laid, the old dealer began dealing. His wrists moved with precision, his slightly gnarled fingers working the cards with an expert's ease. Semyon had no doubt that his shuffle would be as precise as his deal.

To Semyon's pleasant surprise, he was able to pick up a sequence in the first hand, three cards leading to an almighty ace. He knew that he didn't need to signal Victor next to him, or Allie and Owen to Victor's left, because by now they worked together like a well-oiled machine. They had all marked the cards in their minds, put together a story using the signifying words that signaled each card, and were ready to track the sequence when the shuffle finally came. It was just a matter of playing out the deck—

Suddenly, without warning, Semyon felt a stranger's hand on his shoulder.

He knew instinctively what it meant because there was really only one reason why a stranger in a casino would put a hand on your shoulder.

Semyon kept his face steady as he turned in his chair. There were two men in stiff blue suits standing behind him. Neither looked particularly menacing; in fact, the expressions on their faces made them seem like two obsequious waiters about to apologize for a very bad meal. But their name tags dispelled any thoughts of that nature; both were casino supervisors, the English equivalents to pit bosses.

"I'm very sorry for the inconvenience," the closer of the two said, speaking loud enough for the entire team to hear. "But we're not going to be able to let you play at this time."

Semyon was immediately taken aback by the man's politeness. He had expected harsh words, an invitation to a back room, maybe even an accusation of some sort. But this was definitely something different.

Semyon glanced at Victor, who shrugged at him. Semyon raised an eyebrow at the strange English pit boss.

"Might I ask why?"

The man sighed, acting as though it were actually embarrassing for him to continue.

"Well, not having many American guests in our establishment, we faxed your photos on to an associate in the States. He confirmed what we feared, that you're much too skilled for us. Now don't get me wrong; it's absolutely an honor having you at the Ritz. But we simply can't let you play. We'd be happy to buy you dinner, if you're so inclined. Perhaps at one of the nice restaurants across Trafalgar Square? Thanks so much for understanding."

It was like a British Airways advertisement come to life. The man was being unbelievably polite—but he was barring them from the casino. *He had faxed their photos to an associate in the States.* Semyon had a sinking feeling he knew exactly who that associate might be.

Semyon glanced again at Victor, who was already rising to his feet. The bottom line was, they were being kicked out. In the politest way possible. No guns on a golf course, no handcuffs or back rooms. Just a pleasant "you're much too skilled for us."

Well, at least the English had the decency to say it like it was, upfront and honest.

Back outside on the street, Semyon huddled with his team beneath a striped vinyl awning, watching the rain. Victor was the first to speak.

"Well, at least they were nice about it."

The understatement of the evening. It was the kindest barring Semyon had ever heard of. Still, there was no doubt in his mind that Jack Galen had had a hand in it. Fuck, it seemed like the bastard was everywhere. Somehow keeping his eye on them, even thousands of miles away. The thought made Semyon more than a little nervous. In London, they'd been polite about it. But what would happen if the same situation went down in Monte Carlo?

"Victor," Semyon said, and he could tell both Owen and Allie were hanging on his words. "Maybe we should see this as a sign. Maybe we should take our gains and get back to Boston. We've done really well—"

"Are you kidding?" Victor blurted. It was the first time Semyon had ever heard him raise his voice. Even though it was barely perceptible, it was shocking, because Victor was always in control. "We set out to do something, and we're going to do it. You want to turn back just because we got kicked out of a casino? We've been kicked out before. We'll be kicked out again."

"Yeah," Owen interrupted. "But this is different. We're not in Vegas. Things could happen."

"Nothing's going to happen," Victor responded, as if chiding a child. "We're going to hit Monte Carlo, we're going to win, then we're going to go home. And everyone's going to be talking about us for years to come. We're going to be the kids who took down Monte Carlo. Come on, guys, don't go chicken on me."

Semyon bit down on his lower lip. He didn't like Victor's uncharacteristic tone. He knew, inside, that Monte Carlo was going to be more dangerous than anything they'd done before.

"Victor, we should just think this through."

Victor looked at him. The expression on his face wasn't anger, it was—disappointment. He was disappointed in his partner. Semyon was supposed to be cheerleading, not spearheading the opposition.

"I've already thought it through," Victor said, his lips tightening. "That's why I put this team together in the first place. Not just to make money—to make history. I'm hitting Monte Carlo, with or without you guys."

With that, he stepped out from under the awning, into the rain, and headed toward a nearby cabstand. Semyon and the others watched him walk away, the rain soaking his suit. There was a long moment of silence, which Allie finally broke.

"We've gone this far," she said quietly. Semyon looked at her. She was still staring after Victor, and Semyon knew she wanted to follow. It bothered him, how much sway Victor had over her. He had a feeling that even without Semyon, she'd probably go on.

Semyon shifted his gaze to Owen. Owen wasn't looking after Victor, he was just waiting for Semyon to make the call. Because Owen, deep down, was loyal. Not to the team, not to Victor—to Semyon.

Semyon ran a hand over his stubbled jaw.

Christ. Semyon hoped that Victor knew what he was doing. Monte Carlo wasn't Vegas.

Monte Carlo was a different animal entirely.

CHAPTER 29

Monte Carlo

PRESENT DAY

The BMW was sleek and black and moving way too fast, taking the serpentine turns so hard that I was slamming back and forth against the tinted passenger window. The driver had informed me, as I'd first gotten into the special-edition sedan, that the windows—along with the rest of the car's chassis—were bulletproof, and could withstand a direct hit, at close range, from a .45-caliber revolver. Glancing over the edge of one of the hairpin turns, at the thousand-foot drop down to the azure blue Mediterranean, where the waves crashed upward in great, foamy talons against boulder-size rocks, I didn't feel comforted by the fact. A bullet, we could handle, but a slick of oil on the narrow, single-laned road that led up the cliffs to the daunting mountains that protected Monte Carlo from the rest of the world—and specifically, France—would certainly lead to our violent and painful deaths.

Instead of contemplating the long drop down to the sea, I did my best to keep my attention on the driver. Although, in a lot of ways, I found him even more ominous than the suicide-inspiring cliffs. With his platinum blond hair, icy blue eyes, weathered features, and utterly stiff demeanor, he looked like a character right out of a World War II movie; and not one of the smiling, cigar chomping GIs out to save the world, but one of the guys from the other side of the war, the ones with silver skulls on their uniforms, the ones who always seemed to be wearing black leather gloves. Coincidentally, Lucas Anderson was indeed wearing black leather gloves as he gripped the steering wheel with both hands, but I didn't see any skulls on his designer lapels. Thankfully, as far as I knew, Armani hadn't yet come out with the silver skull accessories that would have made Anderson's appearance complete.

He seemed finally to notice the way I was gripping the dashboard with each turn, and he let up a bit on the gas. Or maybe he was just slowing to point out another incredibly magnificent mansion, perched about a hundred yards above the road on a decidedly precarious rocky precipice.

"That belongs to one of my clients as well. Twelve thousand square feet. It has two pools and a staff of seven. Not including the security. It can't be reached from this side of the mountain, only through a twenty-foot electric gate on the far side. Very secure."

Anderson's accent went with his appearance, though I knew that it was Dutch and not German. When he'd first introduced himself to me, he'd described himself as a true resident of Monte Carlo, though in further conversation, he'd finally admitted that he'd been in the city-state only ten years and had spent most of his life commuting between Holland and New York.

"And that one," Anderson continued, pointing up through the windshield toward a Victorian castle built into the sloping cliff above the next hairpin turn. "That's Paul Allen's place. You know, the computer billionaire? It's on the end of a private road with a guardhouse, I think twenty armed guards or so. Very nice place. If it comes on the market, I'll be showing it to all my clients."

I felt my stomach churning as we took a hill at near top speed, then began another ascent. Still, the view outside the window was certainly stunning, and not just the incredible mansions we were touring by at nearly ninety miles per hour. The mountains were lush with greenery and bristling with incredible views, of the Mediterranean down below and also of the city itself, which was perched right above the water on ancient cliffs. From this distance, Monte Carlo appeared like something out of a fairy tale; and having spent two days already touring the immaculate streets, gawking at the lavish architecture, and window shopping at the high-end stores that lined every inch of the city's real estate, I knew that the fairy-tale aura of the place only grew the closer you moved to the city's center.

It was really a place of European legend, the capital of a country that defined romance and wealth, one of the few remaining true monarchies in the world. A place of real live princes and castles and European playboys, of almost unbelievable riches, of order as only a true monarchy can maintain. Monte Carlo was perhaps the wealthiest and arguably the most beautiful city in the world; a residential average per capita income in the tens of millions, with weather so good it made the rest of the world blush—nearly three hundred sunny days a year, never too hot, never cold—with streets lined with Ferraris and Porsches, almost sterile-clean streets that turned into a bizarre racetrack once a year when the city hosted the party of parties, the Formula One Grand Prix, with a harbor so crowded with megayachts it looked as though you could easily walk from deck to deck.

"Yes, very nice views," Anderson said, waving in the direction I was looking. Then he pointed to another one of the houses, this one literally on the peak of a craggy outcropping, with no noticeable road attaching it to the rest of the mountain. "That one is owned by a French casino entrepreneur. I think he bought it for twenty-five million. It can only be reached by funicular. Not even by helicopter. The funicular is operated by four armed guards. Perfectly secure."

I nodded, still keeping my thoughts to myself. It had been like this since we'd left the Hermitage hotel in the center of the city, where we had first met that very morning. Him driving, lecturing me on the various mansions we passed, always noting the secure nature of each of the pieces of property, always making sure I knew how many guards each house employed or how much money had been spent on security systems, trained dogs, funiculars, and hidden helipads. At first, I'd thought there had been some sort of miscommunication; we'd met when he'd approached me after a lecture I'd given to a group of European hedge-fund managers about how gambling relates to the i-banking world, and somewhere in my presentation I'd mentioned some of the dangers involved with taking on Vegas in the way that Semyon and his MIT crew had done. I'd thought that perhaps Anderson had thought that I was in some sort of personal danger because of my work, and would enjoy hearing about the sort of security measures available to the high-end real-estate customer in Monte Carlo. But over the drive, I'd realized that Anderson's pitch was practiced, that this was the tour he gave to all those he considered prospective clients.

Anderson was, in simple terms, a real-estate agent. He sold houses in and around Monte Carlo to the phenomenally wealthy and had done so for nearly ten years. In that time, he'd sold property to royalty, to billionaires, to movie stars, to Grand Prix drivers, and, as he'd pointed out to me again and again, to men whose means to wealth were more obscure—in laymen's terms, the sort of men you didn't ask too many question of, if you valued your health. The sort of men who needed houses that could be reached only by funicular.

He was giving me the guided tour, not because I'd implied that I was interested in buying property in Monte Carlo, but because I'd made comments in my presentation that had led him to think that I knew the sort of people who might be; people who had made enormous amounts of money at young ages, who were the sorts of adventurous souls who'd seek out a fairy-tale kingdom like Monaco.

"Gamblers," Anderson had said to me, when he'd first approached me in the hotel ballroom where I'd given my speech, "have a special affinity for Monte Carlo, because of course all of this started with the casino."

I knew the history, of course, having researched it when I was first invited to speak at the hedge-fund convention. Monte Carlo had indeed grown into the wealthy city it had become because of gambling. In 1863, the royal family had actively decided to make the city into a playland of the rich and powerful by commissioning the greatest casino in the world. The goal was to set Monaco apart from the rest of Europe, and specifically France, which bordered the tiny country on all sides, by turning it into a haven for the ultimate European jet set. The plan had worked beautifully, and over the next century, the city grew in wealth and power, first on the basis of that famous casino, then through the implementation of incredibly lax banking and tax laws. By the time King Ranier III and his son, Prince Albert, took the reins of the monarchy, Monte Carlo had firmly taken its place as a kingdom apart. The ten-minute helicopter ride or forty-minute winding drive from the airport in Nice was like ascending a gilded stairway to a children's-book fantasyland. Except this fantasyland had been designed to satisfy Europe's wealthiest men and women, which meant that, in reality, it had to keep up with the times, it had to cater to the needs and wants of the elite, and it had to be safe.

I'd read somewhere that there were more hidden cameras on the streets of Monte Carlo than in any other city in the world. That the police force almost outnumbered the actual inhabitants, that the city was so secure that there was almost no crime—or at least, no crime that was reported, that made it into the press in the outside world. Monte Carlo was considered one of the safest places on earth, which I guessed was a necessity when you were trying to sell real estate to billionaires.

I noticed that the BMW had finally slowed to a reasonable speed as we took a gentle curve. Then we were pulled into a private driveway, lined on either side by spruce trees.

"I want to show you the front of this house," Anderson was saying as he navigated the BMW up the winding paved driveway. No house had come into view yet, but I assumed there had to be something pretty magnificent around the next corner. "It's all original stone from Athens, flown here by helicopter—"

He stopped talking. Up ahead, there were three men jogging down the private road toward us. They were wearing unidentifiable uniforms—and had automatic rifles slung around their chests. One of the men was waving his finger while shouting at us in French. Anderson glanced at me, then quickly put the car into reverse and sped out of the driveway, back onto the road.

"I guess not today," he said, laughing nervously. "I thought the monsieur was out of town, but he must have returned early. His family is one of the richest in Paris, but he's lived here above Monte Carlo for seven years."

I glanced back over my shoulder as we picked up speed. The three men were standing at the mouth of the private driveway, watching us go. One of them had his hand resting lightly on the grip of his rifle; not pointing it at us, exactly, but menacing just the same.

"You know," I said, finally bringing up the thought that had been eating at me since we'd left the Hermitage, "for what's supposed to be the safest city in the world, these houses have an awful lot of security."

Anderson nodded vigorously.

"Monte Carlo *is* the safest city in the world. But the people who own these houses, some of them are the most dangerous people in the world. So you see the need for balance."

I raised my eyebrows. Balance—an interesting word choice. Men with guns and guarded funiculars, streets bristling with hidden cameras, and a police force that answered to a monarch. A place of immense wealth, a playland for Europe's richest people. A true monarchy with a real live prince and an overwhelming aura of security. A place populated by some of the most dangerous and shadowy people on earth.

Yes, it was a balance.

"A balance of extremes," I said.

"Indeed," Anderson said, and he gave me what seemed to be a wicked little smile. "And it's a balance that one does not want to upset."

I wondered what it would take to upset such a balance. And what would happen if you did.

Then I thought back to Semyon, to his small group of MIT geniuses who dared to try to challenge the most famous casino in the world, the centerpiece of the city itself. A group of MIT masterminds with a system never seen before, trying to take Monte Carlo's Casino de Paris for millions of dollars.

I wondered if Anderson would consider that upsetting the balance of extremes. And I wondered if I told Anderson about Semyon and his team, about what had happened to them in Monte Carlo, would he have been surprised?

I looked at the way his black leather gloves gripped the steering wheel of the bulletproof BMW sedan, at the way he kept glancing in the rearview mirror at the three men with the automatic rifles, receding as we moved higher and higher into the mountains.

I doubted Anderson would have been surprised by anything, after ten years in this beautiful, secure, and terrifying place.

CHAPTER 30

Casino de Paris
Monte Carlo

erraris, Lamborghinis, Porsches, and Rolls-Royces lined up around the semicircular driveway like strange, brightly colored carnivores guarding a watering hole, their sleek curves glistening in the light of a dozen flickering lampposts. The clink of glasses and silverware from the sprawling outdoor Café de Paris, melding with the soft chords of a violin quartet. The breeze, tasting faintly of salt and water, blowing gently upward from the Mediterranean. And above it all, rising up like a gothic movie set, the magnificent Monte Carlo Casino, an architectural masterpiece with frescoes in the style of Boucher, bristling with stone sculptures and bas reliefs, reminiscent of the Paris Opera because indeed the two buildings had been designed by the same man.

Semyon, Victor, Allie, and Owen stood at the foot of the stone steps that led up to the casino, taking in the moment. To their right, the sprawling outdoor café was bustling with tourists, and every few seconds the flash of a camera blinked through the night air. To their left was the magnificent Hotel de Paris, favorite of European royalty and Hollywood movie stars alike, built by the monarch's family. Behind the hotel was a cliff that led all the way down to the sea, and every now and then, they could hear the bells and horns of the many fabulous yachts that were parked at the cliff's base, the toys of the truly wealthy.

Semyon took a deep breath, preparing himself. There were tourists strolling everywhere, but not the sort of tourists you saw in Vegas or London or Amsterdam; the men wore jackets and ties, the women elegant gowns. The couples, hand in hand, seemed all picture-perfect, the men tall and dashing, the women beautiful; their perfection enhanced the movie-set feel of the place. Even the few families Semyon saw admiring the fancy cars parked in front of the casino seemed to have been drawn right out of Central Casting; little blond boys in blazers, little girls with bows in their hair.

"This is what a fairy tale looks like," Owen whispered, and Semyon nodded. He turned his attention back to the casino, to its rising stone facade. The place was actually split into two parts, one housing the Monte Carlo Opera, one the most famous of all gambling palaces. There were two sets of ornate wooden doors leading to the shared lobby, and two uniformed greeters holding the doors open, wide enough so that Semyon could get a good look at the opulence inside, at the marble-and-gold circular atrium, lined with twenty-eight matching onyx Ionic columns, lit from above by gold-and-crystal chandeliers.

"I just don't remember any fairy tales that included four MIT kids," Semyon whispered back. He pulled nervously at the collar of his tuxedo jacket. It was the same rental that he'd worn to the wedding; he'd forfeited the deposit and just kept the damn thing. Owen's tuxedo was also a rental, though it seemed to fit him a little poorly around the shoulders. Victor's Armani tux, was, of course, perfect. And Allie, in a long black dress with a sheer back and a diving neckline, was a vision of pure beauty.

"That fairy tale's still being written," Victor said, then abruptly went up the stone steps. "Okay, team. Let's get this started."

Allie was the first to follow, her high heels clinking against the stone. Semyon glanced at Owen, who smiled weakly and shrugged.

"I was expecting something more literary, you know, like *Braveheart* or *Gladiator* or fuck, even *Patton*. I guess 'let's get this started' will have to do."

After passing through the lavish atrium, they slowly worked their way into the interior of the building. The casino was actually split up into multiple gambling rooms. At first glance, it didn't really look like any casino Semyon had ever seen; it looked more like a palace or even a church: thick Oriental carpets, stained-glass windows, bronze sconces, oil paintings depicting nineteenth-century scenery, and of course, more chandeliers, so many goddamn chandeliers. But then he saw the table games, placed almost haphazardly about the vast rooms, he heard the riff of cards and the spin of roulette wheels— and he began to feel more at home.

It took them a while to find the twenty-one tables, which were in a separate area at the end of a long, sculpture-lined hallway. Semyon counted eight blackjack felts, all staffed by young men in matching dark jackets and ties. There were a fair number of civilians mulling about the tables, though because of the dress code, it was hard to tell how many of them were actually there to gamble and how many had wandered over from the opera for a bit of vicarious fun. Semyon also counted at least six men he identified as casino personnel; though they weren't wearing uniforms, they had buttons on their lapels— though Semyon could have picked them out just from the way they moved. A little too stiff, a little too proper, their eyes scanning the action, flicking quickly here and there.

This wasn't going to be easy, that was for sure. Not only was the room fairly crowded, but the tables were pressed very close together—so close that in some places they looked like they were touching. The minute Semyon's team started winning big, everyone in the room would know about it, and all eyes would be on them. In Vegas, that was usually a good thing, but here, Semyon wasn't so sure.

He felt Victor's hand touch his wrist, and he shifted his attention in the direction Victor was facing. One of the blackjack tables was emptier than the rest, just a man and a woman sitting next to each other in the first- and second-base seats. They looked to be in their midforties, definitely European, the man with slicked-back blond hair and a pair of tiny reading glasses, the woman with her frosted hair bunched up on her head and ruffles in her dark gray gown. Semyon shifted his gaze to the dealer; the young man was waifish, with pale features, thin limbs, and skeletal fingers. He moved deftly—but

Semyon had no doubt that he'd have trouble covering up the back card under the right sort of pressure.

It looked good, as good as they were going to get. As they moved closer, Semyon saw why the table was emptier than the rest; the betting minimum—in French francs—was the equivalent to a thousand U.S. dollars. From what he'd read about the casino, Semyon did not think there was a table maximum; which meant, if things went well, their profit potential was almost unimaginable.

Semyon let Victor take the lead as they reached the table. He could tell from the way Victor threw himself onto the third-base seat that the guru was itching to get into the action.

Victor waited until Allie had taken the seat next to him, draping an arm over his shoulders, before he turned back and suddenly shouted in what Semyon guessed was supposed to be a loud Southwestern accent: "Hey, guys, get yer asses over here. I finally found a goddamn table with a Texas-style minimum bet."

Suddenly every face in the room was turned toward them. Semyon quickly found his voice.

"Well, that's great, Jessy. I got some of these French dollars just dying to get out and get some air."

Semyon knew that in Vegas, the act would have gone down like a lead balloon. His fake accent was even worse than Victor's, a strange mix of Texas and Russia that any American would see right through. But as Victor had assumed, there weren't any Americans in the room, at least that Semyon could see. The expressions on the faces of the nearby patrons weren't of disbelief—just disgust. Which was exactly what Victor was counting on.

Without a word, the elegant couple in the first- and second-base seats gathered up their chips and moved to a different table. Semyon grinned inwardly as he and Owen happily took their places.

"Well now, ain't that nice," Allie said—and finally one of them actually got the accent right. "Them making room for us, and all. First sign of decent manners I've seen all week."

The dealer was smiling at her, and Semyon knew it had much more to do with the way her dress opened up down the front, revealing the smooth, rounded tops of her breasts, than the way she was talking, but at least it kept him busy while Semyon, Victor, and Owen quickly piled their cash onto the table. When the dealer finally turned his attention to the money, his smile didn't change. He simply scooped up the bills and began doling out chips of various colors. Semyon guessed that in this place, half a million dollars was not such a big deal.

Hell, maybe Victor had been right; with no table maximums, there was no telling how much they could win. And so far, Semyon had detected no signs of heat of any kind. None of the men with buttons on their lapels had even given them a second look. The only people who seemed to be watching them were the gamblers at the next table—and that made sense, considering that the next table was so close Semyon could have thrown a few of his chips onto its felt.

Keeping in character, though Semyon wasn't a hundred percent sure what that character was since Victor was in full improv mode—maybe a Texas oil millionaire of some sort—he smiled widely at the gawking Europeans. There were five of them at the next felt: a pretty young woman, with jet-black hair and tan, Italian features; a young man, probably her boyfriend, wearing sunglasses even though it was the middle of the night; an older gentleman, with curls of white hair on his head and a thick beard; and then two more men, in the first-and second-base seats directly across from Semyon, wearing similar gray suits. The two men looked French, and both had militaristic crew cuts and sharp, angled features. Maybe they were brothers; they certainly looked related, not just physically, but in the way they hovered over their cards, only barely registering the loud, obnoxious scene at the nearby table that Victor had created.

Semyon shifted his attention back to his own table as the dealer finished shuffling up the cards. Just in time to watch the young waifish man roll the pile over. Just in time to clearly see a flash of card between those spider-leg fingers, a flash of color, because that bottom card was a jack.

Yes, maybe Victor was right, he thought to himself as the dealer held out the plastic cut card.

They were going to bust Monte Carlo just like they'd busted Vegas. They were going to bust it wide open.

Two hours later, Semyon was straddling his first-base seat like it was a bucking horse, his entire body trembling as the nerves in his skin fired off, a symphony of adrenaline-borne sensation. The cards in front of him were good, a pair of tens, but that had nothing to do with the thrill moving through him, nothing to do with the hairs standing up on the back of his neck. The cards were irrelevant—especially compared with the huge piles of colored chips that were stacked in front of all four players at the table.

Semyon could hardly believe their fortune. They were now up more than a million dollars. A million goddamn dollars. More than any card-counting team in history had won in a single session, more than any card-counting team could hope to win in months—perhaps years—of play. They'd busted the waifish dealer a total of fifteen times. They'd cut to aces nine times. And they'd caught six separate ace sequences. They were absolutely on fire, playing like they were possessed. None of them had missed a single cut, not even by a card. They were killing Monte Carlo. And they weren't finished.

At the moment, there were six bets of twenty thousand dollars on the table. The dealer was showing a four. And Semyon knew, for a fact, that the third card in the deck still to be dealt was a king of spades. He knew this, with certainty, because he'd seen the king at the back of the dealer's roll; he'd cut to exactly the fifty-second card from the bottom. He'd counted down, signaling Victor and the rest of them as they'd gotten closer and closer. And he'd manipulated the hands so that the king was there, waiting and ready.

In a moment, almost certainly, they were going to bust the dealer again, for the sixteenth time. And they were going to add a hundred and twenty thousand dollars to their winnings.

Semyon could not control the grin that was working its way across his lips. He could feel the eyes of the other patrons on him—

and he liked the feeling. For the past twenty minutes, the next table over had barely been playing at all; they'd just been watching the crazy Americans with their loud voices and almost indecipherable accents as they won more and more. Though at first Semyon had been nervous about the proximity of the other table, he had to admit that he loved having an audience. Let these stiff fucking Europeans see what a bunch of geeky MIT kids could do.

He glanced over his shoulder, just to see their faces. To his surprise, the Italian woman was smiling back at him. Her boyfriend wasn't smiling, but he'd finally taken off his sunglasses, perhaps to get a better look at what was going on at Semyon's table. The elderly, white-haired gentleman had left, but a woman with frizzy blond hair had taken his place, and she, too, was watching the display. Only the two Frenchmen at the end of the table weren't really looking, seemingly more interested in their own cards, their own measly three-hundred-dollar bets. One of them was in the midst of splitting a pair of eights, adding chips from his stack. The other was tapping his stubby fingers against the felt and only incidentally glanced in Semyon's direction—then quickly turned back to his own cards.

Semyon paused.

Something suddenly bothered him about the two Frenchmen. He couldn't quite put his finger on it, but something wasn't right about them. He looked at their crew cuts, at the way they were concentrating on their hands, at their gray suits—and he had a strange thought. They didn't look like regular three-hundred-dollar players out for a night on the town. They looked like—well, they looked like Owen and Semyon, playing roles. Their real intentions were something else entirely.

Semyon quickly turned back to his own table. His heart was now pounding in his chest. His excitement from the moment before had vanished, and now the trembling in his body had nothing to do with the money on the table.

Fuck, was he overreacting? Was he just being paranoid?

He glanced at Victor. He could tell by his smile that all Victor was thinking about was that coming king. Victor hadn't noticed the two Frenchmen.

The dealer was now facing Semyon, asking if he wanted to hit or

250 ♦ Ben Mezrich

stay. Semyon waved his hand, almost out of reflex. He didn't care about the cards anymore. He just wanted to get out of there.

"Victor," he said, under his breath. But Victor ignored him, taking one card on his first hand, which was a fourteen. He drew a three for a hard seventeen. Then he took a second card on a nine, drawing a ten for a nineteen. The deal moved to Owen, who had a twelve. Owen took a card, busting. Now the king was the next one to come out. If Allie held—which of course she would, since she also knew that the king was coming—then they'd bust the dealer for sure. He'd have to pay the table.

"Victor," Semyon said again, louder now. "We need to leave. Now."

Victor looked at him like he was crazy. He saw that Semyon was serious, and for a brief second, he seemed to consider what that meant. Then he shook his head.

"In a minute," Victor said. He was waiting for the dealer to turn over the card. But the dealer had paused, straightening up the chips in his rack. Semyon had seen the tactic before. The dealer was stalling. Again, maybe he was being paranoid—but he didn't think so. He glanced back at the two Frenchmen. They were still sitting at the other table—but now they were both glancing at Semyon, then quickly looking away.

Shit, this was bad.

"Now," Semyon gritted. "We need to get up and leave right now."

"One more card," Victor gritted back.

Christ, Victor had lost it. He was so intent on busting the casino that he wasn't thinking straight anymore. Semyon had to take charge. Because he had a feeling this was going to get ugly, fast.

The dealer finally finished with the chips and moved his hand slowly toward the shoe, as if to deal the next card. But Semyon could see it in his eyes—something had changed. That card was never going to come.

Semyon rose from his seat, grabbing at his pile of chips. Owen and Allie saw him move and, without thinking, were both up on their feet. Victor noticed what was happening, saw that Semyon was serious, and finally snapped back to reality. His hands flashed over his chips and then he, too, was up, shoving the chips into his pockets. Semyon turned, ready to run—

And realized that it was too late.

The two Frenchmen with the crew cuts were coming toward him from the right, moving fast. Four more men with buttons on their lapels were coming in from the left. And six more were directly in front, clearing a path through the gawking European gamblers.

Christ.

"Hey, what's this all about . . . ?" Victor shouted, but his voice trailed off as the men descended on them.

Suddenly Semyon felt himself being spun around, his arms yanked down behind his back. Cold metal touched his wrists, and he realized that a pair of handcuffs had been placed on them. He was shoved forward, and saw that Victor had also been cuffed and was being pushed next to him. Owen had three of the men on him, two from behind, one in front. Then Semyon heard Allie's voice, and he yanked his head to the side. Two men had her by the arms, roughly trying to pin them behind her back.

Without thinking, Semyon started struggling, trying to get to Allie, to somehow help her. But the hands on his arms were too strong. He felt something hard pressed against his throat—and realized it was a police baton. For a brief second, he couldn't breathe, and his eyes spun back in his head. Then the baton loosened, and he gasped.

"Remember where you are," a heavy French accent whispered in his ear.

Remember where you are.

○ ◎

The next hour was a complete and brutal blur.

The four of them, bodily dragged—by no fewer than twelve men in jackets and ties—out the front doors of the famous casino and down the stone steps to a waiting police van. The van racing out past the Ferraris and Lamborghinis, the four of them cuffed in the back, jammed next to one another on hard steel benches that swayed back and forth as the vehicle wound down the narrow European roads. None of them saying anything because they weren't alone; the two Frenchmen with the crew cuts—policemen, it turned out—were in the back with them, staring with hard, cold eyes. The van descending farther and farther, seemingly forever, the world jerking up and down with the ancient cobblestone roads, and always, those damn Frenchmen with their staring dead eyes.

Finally, the van jerked to a stop. The rear doors were flung open, and the four of them were dragged outside, then led through a back door into what looked to be some sort of police station, cut into the base of a high cliff. Down a long hallway to a flight of stone stairs, then another hallway, then more stairs. They found themselves being led down a stone tunnel lined with steel doors. Then, one at a time, they were separated from the group. Each one of them led to a different steel door.

And that's when the real fun began.

T he cell was a perfect windowless cube, maybe twelve feet across, with sheer stone walls and a high ceiling. The only light in the room came from a single, bare lightbulb, hanging from a twist of exposed wire. There was no furniture in the room, not even a chair. Just a drain in the center of the floor. A fucking drain.

Semyon was sitting with his back to one wall, facing the steel door. He wasn't sure how long he'd been sitting there, or how long it had been since the Frenchman had left him alone.

He hated that man, more than anyone he'd ever met in his life. Not because of anything the man had done—but because of the way he'd looked at him, the disdain in his cold eyes.

The interrogation had started almost the moment the four of them had been separated and led to their own cells. Semyon had been followed inside by the Frenchman, the larger of the crew-cutted duo he had first spotted in the casino. The man had pointed to the floor, and Semyon had dutifully lowered himself into a sitting position against the cold cement. Then the man had started in with the questions.

"Who else are you working with?"

"How many of you are there?"

"How much money have you won?"

"How were you winning?"

"What is your system?"

Semyon had kept his mouth shut. The man's face had gone redder and redder. His thin lips had nearly vanished against his jagged, yellow teeth. And still, he kept on.

"What is your system?"

"What is your system?"

Semyon wasn't sure where he had gotten the strength to resist the questioning, as scared as he was, as terrifying as the whole situation was. But deep down, he was more afraid of answering than of remaining silent. Because if he had told the man what he wanted to hear—that his famous casino was beatable, that Victor had cracked the game of blackjack in a way that nobody else had done before, that the three techniques could truly beat the house—how would he react? Would he want to make an example of them? Would he want to silence them? How would he react?

So Semyon had just sat there, staring at the man's face.

Eventually the man simply got tired of asking the questions. He crossed his arms against his chest and bared his yellow teeth.

"Card counting is illegal here," he said in that thick accent, although he had to know that Semyon and his team were not card counters, or else all of this wouldn't have happened; they'd simply have been thrown out like every other team that ever had tried to beat the house. "This is not America. We can put you in jail for a very long time. Your parents will never see you again. You will never see your friends again. Is that what you want?"

Semyon lowered his eyes.

"Did you think you'd get away with it? Did you think we wouldn't check the photos of four Americans betting big with our friends in London and Las Vegas?"

Semyon blinked. Jack Galen again. They should have known. After what had happened in London, they should have turned around and gone home.

The Frenchman gave him another minute to think about it, then turned and left the cell, slamming the steel door shut behind him.

Finally, Semyon was alone, with his thoughts and that damn drain in the middle of the floor. Sitting there, back against the wall, thinking about Victor, that fuck, who couldn't quit while he was ahead, and Owen, probably getting his ass kicked in an adjoining cell for being an impudent American brat, and Allie.

God, Allie. Semyon closed his eyes. He hoped that she was okay. He had never felt this helpless before in his life. He didn't really care what they did to him—he could handle the intimidation, the physical violence, the fear—but he prayed that they just left her alone.

Because if something happened to her—he wasn't sure what he'd do.

S emyon was half asleep, leaning against that stone wall, when they finally came for him again.

He didn't resist as they lifted him back to his feet, led him back out into the hallway. Didn't resist as they dragged him back up the stairs, through the police station. Didn't resist as they led him back out to the waiting van.

When he was pushed inside, he saw that Victor, Owen, and Allie were already sitting on the metal bench, and his body filled with relief. There were bruises all over Owen's face, and Victor's hair was sticking straight up, his suit ripped and hanging off him like the tattered clothes of a scarecrow—but Allie looked like she hadn't been touched. Her face was pale, her eyes bloodshot—but she didn't look harmed.

Semyon lowered himself next to her, and she leaned against him. Her body was shaking, and he tried to calm her with a weak smile. He was glad that it was only the four of them in the back of the van because now they could finally talk.

"It's going to be okay," he said, though deep down, he really wasn't sure. "If they'd wanted to hurt us, they'd have done so already. I think they're going to let us go."

"I tried asking for the American ambassador," Allie said. "They just laughed at me."

"You're lucky," Owen said, somehow grinning. "When I asked them for a cigarette, they played Ping-Pong with my face."

For some reason, that got them all laughing. Maybe it was just a way to break the tension. The van was moving fast now, obviously on some sort of highway. Semyon looked at Victor, but Victor's face was unreadable. Maybe he was sorry that he had led them into this mess—and maybe he wasn't. Or maybe he was just thinking about the million dollars they had won—most of which had been left on the table. Semyon still had some chips in his pockets, perhaps more than a hundred thousand—but he didn't think he'd be cashing them in anytime soon.

"Well," Victor finally said, because he had to say something. "That didn't go exactly as planned."

Semyon had the sudden urge to punch Victor in his delicate face. It was probably a good thing his hands were still cuffed behind his back.

Without warning, the van started to slow, then came to an abrupt stop. There were footsteps coming around both sides—and then the doors were pulled open. The two Frenchmen were standing outside—and both had guns in their hands. The bigger of the two, the same one with the yellow teeth who had interrogated Semyon, gestured with his free hand, indicating that they were to get out of the van.

Semyon climbed out first, followed by Victor and the rest. The van was parked on the side of some sort of highway. They had just passed through a tunnel, the dark mouth of which gaped at them. In the other direction, the highway seemed to lead on forever. It was still dark out—dawn was probably an hour or two away—and Semyon couldn't see any lights in the near distance. They were probably miles from anything resembling a town.

"Over there. In the grass. In a line, with your backs to the road."

The Frenchman's voice was like a leather belt pulled tight. Semyon glanced at Allie, then at Owen and Victor. Victor shrugged. What choice did they have?

A second later, they were shoulder to shoulder in the grass. Semyon's mind was whirling. He was staring into the darkness, listening to the sounds of the two men's shoes as they walked back and forth on the road behind them. It was simply the most terrifying moment of his life. He didn't really think that the two men would kill them—right there, on the side of the road. But then again, he couldn't know for sure.

Remember where you are.

Then, finally, the man with the yellow teeth stopped directly behind Semyon and spoke in his thick accent.

"You are now on French soil. If you ever return to Monte Carlo, if you ever set foot in our casino again, you will be buried here. It's as simple as that. Do you understand?"

The silence was brief and loaded. Then Victor answered for all of them.

"Absolutely."

"Good. His Highness Prince Rainier sends his regards. He takes it quite personally when someone tries to steal from his casino. I wouldn't want to cross a prince twice."

And with that, the two police officers got back in the van and drove off, leaving the four of them standing in the grass at the side of the road.

The next thing Semyon knew, Allie's arms were around him. Then they were all hugging. They were in the middle of nowhere, battered and beaten. But they were still alive.

CHAPTER 31

Boston, Massachusetts

Nearly twenty-four hours later, just as they passed through U.S. passport control and out into the international terminal at Boston's Logan Airport, Semyon realized that their problems were just beginning.

Owen was the first to see the uniformed police officers. Four of them, standing together by the entrance to the baggage area, the lead one, Irish and burly, holding a black-and-white photograph in his thick hands, while his colleagues scanned the crowd that had just come off of the Air France flight from Nice.

"Uh-oh," Owen said, nearly stopping in his tracks. "Someone forget to pay a parking ticket?"

Then Semyon and the rest of the team saw the cops. Semyon's stomach flipped over, and he nearly sat right down on the floor. *Not again.*

It had already been the longest twenty-four hours of his life. First the two-mile hike down the deserted highway in France, until they'd managed to flag down a passing newspaper truck and begged a lift to the airport in Nice. Then the long wait in the airport terminal for the Boston-bound flight, most of which had been spent trying to calculate how much money they had managed to get out of Monte Carlo, in chips and cash. Semyon wasn't sure why the Monte Carlo police hadn't confiscated their winnings; then again, as rich as the little city-state was, he doubted that the display of force had had anything to do with money. Like the Frenchman had said, it had been personal, between the prince and the kids who had busted his casino. In any event, they'd managed to escape with a good portion of their winnings, not the least reason for which was that Owen had been fast with his hands, jamming nearly three hundred thousand into the pockets of his pants. All told, they were up more than a million from their European adventure. A pretty good take, all things considered.

But from the looks of the police officers at the edge of the baggage area, the adventure wasn't over yet. Maybe not by a long shot.

"You think they're looking for us?" Allie asked.

The cops were staring in their direction. One of the officers pointed directly at Victor, and they were all suddenly rushing forward, the big, burly one already reaching for his handcuffs.

"No," Victor said, shocked. "I think they're looking for me."

Before any of them could react, the cops were on him. The burly officer had his cuffs on Victor's wrists while another was reading him his rights. Thankfully, and surprisingly, they didn't seem to care about Semyon, Allie, or Owen. Just Victor.

"Victor Cassius," the burly officer said, after the rights had been served. "You want to make this easy on yourself?"

Victor looked at Semyon, as the cops dragged him through the crowd of gawkers that was now surrounding them, back toward the baggage area, and Semyon could finally read an expression on the young man's delicate face.

Utter dismay. Victor had no idea what he was being arrested for. Semyon knew only one thing—it wasn't about blackjack. They were back on American soil, thankfully, where you couldn't get arrested for playing cards. At least, not as far as he knew.

○ ○

A burst water pipe. Can you believe it?"

Three hours later, he and Allie were alone in the MIT classroom off the Infinite Corridor, the place where it had all begun. She was leaning against the desk at the front of the room, and Semyon was sitting across from her, on one of the institutional-style plastic chairs, his legs stretched out in front of him, his arms hanging limp at his sides. He was exhausted—beyond exhausted—and it felt like the day was never going to end. After they'd followed the squad car to the police station, then waited for almost an hour before Victor's lawyer finally showed up, they'd finally gotten an explanation for the arrest.

"A fucking water pipe," Allie repeated, her hands flat on the desk behind her. "That just goes to show you. All the planning in the world, and the littlest thing can bring you down."

She bent to roll down the cuff of her jeans leg, her T-shirt rising with the motion, showing a sliver of skin at her waist. She'd obviously dressed quickly; after the lawyer had explained things, she and Semyon had agreed to meet in the classroom to discuss the situation, but they'd wanted to return to their respective apartments first to shower and change. Allie's hair was still damp, and Semyon could smell the honeysuckle in her shampoo.

"I don't think it will be too serious," Semyon said, rubbing his tired eyes and wondering what the hell was taking Owen so long. "His lawyer said he thinks he will get off with probation and a stiff fine. He won't do jail time."

"But he'll have a record," Allie said. "That's going to make it that much harder in the casinos."

Semyon knew she was right. Victor's blackjack days were now definitely numbered. And it was true, it was all because of a broken water pipe. It had nothing to do with what had happened in Monte Carlo, had nothing to do with blackjack or Jack Galen. Just a fucking busted water pipe in his basement apartment.

It had happened two days ago, while they were still in Europe. A water pipe running through the ceiling of Victor's apartment had burst, flooding the entire place, as well as the apartment next door. The fire department had been called—and when they'd broken inside of the basement apartment, the first thing they'd seen was the fake-ID machine. The police had then been called and a search conducted. When the police stumbled on nearly two hundred thousand dollars that Victor hadn't yet transferred to the safe in the MIT building, they'd come to the conclusion that he was running a fake-ID operation, selling the plastic cards to underage college kids. What else would explain that kind of money and the ID machine? So they'd filed for the arrest warrant, and the rest was history.

"Semyon," Allie said, finishing with her jeans and stretching her back. "Do you think we should just quit? Take what money we've earned and call it a day?"

Semyon knew that it wasn't just the water pipe that had inspired Allie's thought. It was Monte Carlo, those cement cells in the basement of the police station, that moment when they were standing in the grass by the side of the road, their backs to the two men with guns.

"I don't know," he said. It seemed so foolish, to give up on a system that had made them so much money, a system that was so powerful. Maybe too powerful.

"Were you scared?" Allie asked. Her voice had gone quiet. Now she was definitely talking about Monte Carlo. Semyon looked at her face, at the way her blue eyes had become almost childlike. At the way her lips pressed together. She looked like she wanted to cry. He had never seen her cry. In a lot of ways, she was the strongest girl he'd ever met. But there they were, the tears, building at the corners of her beautiful eyes.

And then Semyon was up and out of his chair. He didn't really know what he was doing, but he wanted to touch her.

He moved to the desk and put his hands on her knees. He could feel her warm skin through the thin material of her jeans. He leaned forward, his face inches from hers. Her head was down, her eyes lowered, and she was trying to hide her emotion. But he didn't want her to hide anything from him. He reached forward, took her chin in his hand, and lifted her face to his. Their lips came together, first soft—then hard, mouths coming open, tongues touching. His hands slid up to her T-shirt, then underneath, up her warm, smooth skin to the bottom of her bra. She fell back against the desk and he was on top of her, no longer thinking, the tension of six months of being by her side, of being so close, of playing out so many roles and sharing so many experiences, all of it exploding inside him as he pressed against her, as his hands found her perfect round breasts, her hard nipples, the jut of her collarbone, then down across her criminally flat stomach, to the waistband of her jeans, then beneath, toward the heat that was coming from below her stomach. Her hands were on him, too, beneath his shirt, then grabbing at his belt.

Suddenly, his hands down her pants, hers down his, he realized where they were. He lifted his mouth from hers, looked toward the closed classroom door.

"Should we—"

"Semyon, just shut the fuck up and kiss me."

All of it vanished, the classroom and the Infinite Corridor, the sense of time and place. All of it vanished in the heat of his hands on her body, his lips pressed hard against hers.

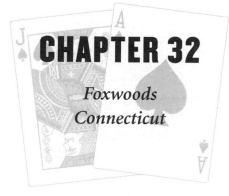

CHAPTER 32

Foxwoods
Connecticut

ell, you've really done it this time. You've really fucking done it.

Owen stared at the card on the table in front of him and felt like he was about to throw up. A two of diamonds. A god-damn two.

He knew, instinctively, what that meant. The card was the second one dealt on his last hand, the third-base seat, the last card before the dealer's up card. The card was supposed to be an ace, because he had spotted that ace as the last card on the rolled and shuffled deck, he'd cut to fifty-three, he'd manipulated and counted down the hands—and that fucking card was supposed to be a fucking ace.

But it was a two. Which meant—

He watched as the dealer pulled the next card from the deck and gave it to himself. He watched as the card landed faceup on the felt. He watched as it almost miraculously became that ace in front of his eyes.

Christ.

Owen had miscalculated. He'd miscounted, and given the ace to the dealer instead of himself. Which meant he was going to lose the three hands he had spread out across the table, all at maximum bet. Thirty thousand more dollars of the team's money. All because he had miscounted by one card, blown the first technique, and given the ace to the dealer.

He shut his eyes and lowered his head toward the felt. The dealer stared at him, but didn't say anything, letting Owen have his moment. The dealer had probably seen it all before. A guy betting too much, losing his shirt. He'd probably seen it so many times, it was a cliché to him. A gambler doing what gamblers do. Thirty thousand dollars, down the drain. Handed right to the casino.

That wasn't the worst of it. In the past hour since he'd arrived at Foxwoods, the Indian resort in Connecticut, Owen was down a total of one hundred and sixty thousand dollars. The ace, given to the dealer, was just the final straw. The one that broke the gambler's back.

How could he have been so stupid?

After Victor's arrest, he'd realized that he still had almost two hundred thousand dollars on him from the flight back from Nice. In the confusion after the arrest, they'd not had time to transfer the money back to the team's safe or to make any arrangements for the transfer to occur.

When they'd separated after the lawyer had explained the situation to them, Semyon and Allie heading to the classroom to discuss things, Owen left on his own. He had felt the money pressing against his back, still in the Velcro pack he'd worn on the plane. He'd suddenly had the thought that maybe he could use that money for something good; to pay back the team for what he'd already lost, the fifty thousand he'd had stolen by the escort, and the other seventy-five he'd lost in Aruba. He wasn't sure where the thought had come from, or why it seemed so possible to him at the time. Maybe after what had happened in Monte Carlo, he had gone a little crazy; that cell with the drain in the floor, how it had reminded him of another cell in another place entirely—but whatever the reason, the thought had quickly taken hold, become a mission.

He'd gone straight to his motorcycle and taken off for the short trip to Foxwoods. His plan had been foolishly simple; cut to some aces, maybe bust the dealer once or twice—and then Semyon and Victor would see that he could be trusted, that he wasn't just a big negative mark on the team, that he was a real contributor. He'd pay them back for all that he'd lost, and maybe more.

He knew that cutting to aces on one's own was hard—but it wasn't impossible. With just a little luck on his side, he could win. He could win big.

But Owen should have known better. He had never had luck on his side. The two of diamonds in his hand, facing the ace in the dealer's hand, said it all. He'd gone from being a drug addict, to being in jail, to barely making it through MIT. Then he'd found Victor's team, he'd found Semyon, and things had for once looked like they were finally going to be better. And then his old ways had come back, he'd fucked up, again and again. And just when he had thought he had it together—he'd fucked up again. Except, this time, there would be no second chances.

Eyes closed, head inches from the blackjack felt, Owen realized that he had hit bottom. Third strike, and he was finally out.

This time, there was no way Semyon was going to forgive and forget.

CHAPTER 33

Boston, Massachusetts

ix days, Victor. It's been six days. I'm telling you, that's not like him."

Semyon stamped his foot on the cement floor, listening to the sound echo through the dank basement. But Victor didn't even look up. He remained exactly where he was, on his knees in front of the open safe, carefully counting the banded stacks of hundreds inside. Because that's what was important to him. Not the fact that Owen had been missing for nearly a week. Not the fact that the team was pretty much in shambles, after what had happened in Monte Carlo and then Victor's arrest. Nothing really mattered, except the banded stacks of hundreds.

"He would have called if something wasn't wrong. He would have at least let me know where he was."

Victor shrugged. The motion pulled at the material of his pale green suit. Semyon wondered if he had worn the same suit at his hearing, early that morning. Victor had easily posted bond, and as his lawyer had said, he probably would get off with just probation and a fine—but his blackjack team was now seriously in jeopardy. Because with a police record, he'd be a real liability in the casinos. By association, he could get them all in big trouble.

"I'm sure he's fine," Victor responded, his voice echoing out of the massive safe. "He's certainly not hurting for cash. He's got two hundred thousand dollars of our money on him."

Semyon didn't like Victor's tone.

"You think he ran off with the money? That's bullshit, Victor. He's made almost that much playing with us, why would he want to steal?"

Victor shrugged again.

"I didn't say he stole it. I just said he's still got it on him. Unless he's fucked up again."

Semyon opened his mouth, but didn't say anything. Victor finally rose from the safe and turned toward him. His face was passive, but his lips were turned down slightly at the corners.

"Yes, Semyon. I know all about the money he lost in Aruba, and the fifty thousand that disappeared in Vegas. You don't think I keep track of every penny on this team? Owen's a loose cannon. I told you that in Atlantic City. He's dangerous."

Semyon shook off his surprise that Victor had known about Owen's losses all along. How could Victor accuse Owen of being dangerous, after what had happened in Monte Carlo? After his own arrest for the ID-making machine? They were all dangerous. They were all part of a scheme that was inherently dangerous.

"But six days, Victor. It's a long time to be gone. And after what happened in Europe, we need to stick together. We need to keep track of each other."

"I know you're keeping good track of Allie, at the moment."

Victor was smiling, but Semyon noticed a little venom in his eyes. He wondered if Victor was jealous. Since their moment in the classroom, Semyon and Allie had been near inseparable. They'd spent the past few nights at her place, sharing her bed. He could still smell her skin on his clothes. In fact, the reason he hadn't really noticed that Owen was gone for so long was that he'd been so caught up in what was going on with Allie. He could hardly believe that finally they were together. He'd never felt so strongly for anyone.

Did Victor have feelings for her, too? No, he doubted that was it. Victor wasn't jealous because he secretly liked Allie. He was jealous because his team was no longer loyal only to him. Other loyalties had sprung up along the way. Semyon and Allie. And Semyon and Owen.

Semyon set his jaw. He knew that Owen had not absconded with the two hundred thousand dollars. He knew, in his heart, that something had happened to his friend. After he'd realized that Owen was gone, he'd gone by his apartment. The landlord had let him in, and Semyon had seen that Owen's duffel was missing, along with some clothes. Owen wouldn't have just left, without telling Semyon, unless something had happened. Something bad.

"Victor, I'm going to find Owen."

Victor's smile vanished. He could hear the determination in Semyon's voice.

"How? You have no idea where he is."

Semyon nodded. He had thought this through.

"I think I know someone who might."

Victor stared at him. The mastermind of the three techniques figured it out almost immediately. The skin of his cheeks seemed to tighten against his cheekbones.

"No fucking way, Semyon. You can't. It will be over for us. You'll have to tell him who we really are."

Semyon shrugged. He knew that his plan was dangerous, and crazy. It could have enormous ramifications for the team. But he also knew that Owen was in trouble. It was more than a feeling. Owen would not simply vanish, not after what they'd been through as a team.

"You and I," Victor said quietly, his little body framed by the enormous safe behind him. "If you do this—we're done."

Images flashed through Semyon's mind. Memories. That first meeting, in the classroom, where he had first set eyes on Victor. His inauguration into the team. The first time he'd successfully busted a dealer. Then darker memories. The plane crash. The back room. Monte Carlo.

He looked at Victor, with the massive safe full of money behind him. Victor would never be a real partner. He'd always be the boss.

Semyon didn't want to have a boss anymore. He had Allie. He didn't need Victor.

He nodded, then turned toward the door.

"Good-bye, Victor."

T he phone felt heavy against Semyon's ear. His mind was spinning as each ring crashed through his skull, as the moment when someone on the other end finally answered drew nearer and nearer and nearer—and then, there it was, a male voice on the other end.

"Jack Galen here. Can I help you?"

Semyon shut his eyes. He was lying flat on his back on the floor of his apartment. There were decks of cards spread out all around him, the chaos of six months of practice, of cutting and shuffling and concentrating. Of everything he had ever learned.

"Hello? Can I help you?"

Semyon took a deep breath, then spoke into the phone.

"Mr. Galen, my name is Semyon Dukach. I've also gone by the name Nikolai Nogov, and Dr. Michael Nussbaum, among many others."

There was a long pause on the other end of the line. Semyon could almost picture Galen dropping heavily into the seat behind his desk. It hadn't been hard to track down Galen's phone number. The Griffin Detective Agency had been happy to provide it, once Semyon had told them who he was, even though Galen was not affiliated with the agency.

Finally, Galen whistled low and found his voice.

"Well, well, well. Mr. Dukach. Back from Europe, I assume."

Semyon wondered if Galen was going to try to trace the phone call. He didn't really care. He knew that to get what he wanted, he was going to have to give up his anonymity. He was going to have to deal with the devil, possibly face-to-face.

"Mr. Galen, I have a problem. One of my teammates is missing. You and I, I know we're on opposite sides of the game, but Mr. Galen, this is about a person's life. My friend—I can't find him. I think maybe you can help me."

Semyon could hear Galen breathing on the other end. He was

thinking it through. Galen was as much of an opportunist as any of them. And the opportunity to catch the biggest thorn in his side had now just presented itself.

"If I do this, Mr. Dukach, you know that you're finished, right? Whatever your system is, you have to promise me that you'll never use it in Las Vegas again. Not as long as I'm on watch. Is it a deal?"

Semyon reached out with a hand, let his fingers leaf through the loose cards on his apartment floor.

"Yes, Mr. Galen. It's a deal. Do you know where my teammate is?"

Semyon could almost see Galen's smile on the other end of the phone.

"I do. But you need to get out here. And Semyon—if you want to see your friend again, you'd better hurry."

CHAPTER 34

The Ranch
Carson City, Nevada

A n uncharacteristically cool wind was blowing across the parking lot as Semyon slowly stepped out of his rented Ford Mustang convertible, quietly shutting the door behind him. The Mustang hadn't been his first choice, seeming a little too flashy for the long and lonely drive out into the desert, but it was all the airport car-rental agency had had available at such short notice. If Semyon had been with Allie, the top down and the night wind blowing as they burned up the two-lane highway, the Mustang would have been perfect. But there was no way in hell Semyon was going to bring Allie with him on this trip to Vegas. Because this trip to Vegas wasn't about blackjack, or about the team, or about what was happening between the two of them. This trip was about Semyon and Owen—and Jack Galen.

Semyon took a deep breath after he spotted the thin, dandyish, British "consultant" on the other side of the parking lot, leaning against a pitch-black Mercedes sedan. Galen looked to be alone, standing there casually next to the car, both hands in his pockets, as the wind pulled at his lavish locks of blond hair.

Semyon approached him quickly, his nerves firing. The sense of urgency had been rising inside of him ever since he'd hung up the phone—now nearly nine hours ago. He had made it to Vegas from Boston in near-record time, but he still didn't know if he had been quick enough. He didn't know what Galen hadn't yet told him: where Owen was, or what had happened to his friend.

Galen looked up as Semyon reached the front of the Mercedes, then smiled that thin, lipless smile.

"Good evening, Semyon. Ever been here before?"

Galen gestured toward his left, toward the low, squatting building on the other side of the parking lot. The first thing Semyon saw was the neon sign—of course, there had to be a goddamn neon sign. Beneath that, the Wild West–style front entrance, separated from the parking lot by a small driveway lined with decorative hitching posts and a high metal gate.

"Never," Semyon said. He'd heard about these brothels, of course; anyone who'd been to Vegas as many times as he had knew about the brothels. He'd had no problem getting directions to the place from the car-rental agent, who'd winked at him as he'd handed across a printed map.

"Well, your friend has. Many times. I can tell by your expression that you didn't know that. But I can also tell that you're not surprised."

Semyon shrugged. Galen was right; he hadn't known that Owen had visited the Ranch before, but he definitely wasn't surprised. However, he couldn't understand why Owen would be there now. Why, so soon after their trip to Europe, he'd want to get on another plane and fly to Vegas.

"Is he in there now?"

Galen nodded.

"He's been there for two days. He paid for a room, then asked to be left alone. Not a usual request in a place like this. But he gave them ten thousand dollars, cash. And he's been locked in there ever since."

Semyon's heart was pounding. Two days alone in a room in a brothel. Semyon turned and started toward the building beneath the neon sign. Galen took a few steps after him.

"Semyon, don't forget our deal."

But Semyon was already at the driveway. He took the last few steps in a near run, tore open the gate, and slammed his fist against the door. Again and again.

The door was buzzed open, and Semyon burst inside. He barely noticed the sea of red velvet, the kitschy Old West decor. He ran straight toward the woman sitting on a stool by the bar, the madam of the house. Short, squat, with a ball of dyed hair on her head.

"Sir, you'll have to wait in the front room—"

"I'm sorry, ma'am. My friend is in trouble. He's here, in one of your rooms. He paid ten thousand dollars and asked to be left alone."

The woman looked Semyon over. She could see the fear in his eyes. She made a decision and nodded. Then she rose from the stool.

"Let me get the key," she said, reaching behind the bar.

A minute later, they were moving down the carpeted hallway that led into the depths of the building. The woman was moving slowly, because of her short, stubby legs, and Semyon wanted to grab the key out of her hand and run, find Owen, stop whatever it is he was doing, get him the fuck back to Boston, get him the help that he probably needed. But Semyon kept his frustration in check, following the woman's slow progress, step by step by step.

"Here we go, Room 232," she said, finally stopping in front of one of the doors. She went to work on the lock with the key. "Usually, one of our girls has to bring you back here, but your friend was so insistent—"

The door came unlocked and Semyon pushed right past her, shoving the door open with his palm. He dove into the room, his eyes moving fast, taking everything in at once. The shag carpeting, the king-size bed, the mirrors on the ceiling, the red curtains in the windows. He stepped farther into the room.

"Owen? Owen, man, where the fuck—"

He saw him. On the floor. Partially obscured by the bed, on the shag carpet right up against the wall. Lying facedown, arms twisted beneath him, wearing only his jeans. Even from that vantage, Semyon could see the yellow rubber tube tied around Owen's right arm, just above the elbow. And the two syringes, both empty, lying on the floor by his head.

Semyon leaped forward.

"My God, Owen!"

Owen didn't move. Semyon dropped to his knees by the bed and grabbed Owen by his shoulder, rolling him onto his back. To his relief, Owen's skin still felt warm. He quickly held a palm over his mouth, felt a slight burst of air against his hand. Then he leaned over his friend's chest, listening for his heart. Faint, but it was there. Owen was still alive. But barely.

Semyon cradled his friend in his arms and looked up at the woman, who was still standing in the doorway, shock in her face.

"A hospital! I have to get him to a hospital!"

She nodded, her bun of hair bobbing up and down. "There's a clinic, three miles down the highway. Opposite the way you came, if you came from Vegas."

Semyon hooked his arms under Owen's body and struggled to his feet. Owen's limbs hung down, limp, making him even heavier. He was completely out, completely fucked-up, like a rubber mannequin. But not dead. Not yet.

Semyon stumbled with Owen's body across the room and back out into the hallway. He nearly fell twice on his way through the red-velvet-lined lobby, but finally made it to the metal gate outside. Then out into the parking lot.

He saw that Galen's Mercedes was gone. In its place, someone had parked an oversize Chevy truck. The truck looked empty; nobody around to help. Semyon kept on moving, step-by-step, back toward his rented Mustang.

He made it to the car and yanked the passenger-side door open. With a grunt, he managed to roll Owen inside. Then he slammed the door shut and rushed to the driver's door. He had just gotten it open when he heard a loud sound from behind him.

He turned, and saw that there were now two men standing in front of the Chevy truck, about ten yards from where Semyon's car was parked. He didn't recognize either of them; big, scruffy-looking men, both in work shirts and jeans. One of the men was holding a baseball bat. As Semyon watched, the man knocked the business end of the bat against the pavement. Then he smiled, and suddenly the two men were coming forward, the bat in the air.

Semyon's eyes went wide. He leaped into the Mustang and slammed the door shut behind him. He jammed the key into the ignition, turned it hard, and heard the engine roar to life. Then he looked back into the rearview mirror.

The man with the bat was right behind the car. The bat was in the air, rising, and in a second it would come down right into the Mustang's rear window.

Semyon didn't pause to think. Owen was next to him in the front seat, dying.

He shoved the car into reverse and slammed his foot down on the gas pedal. The car lurched backward—and there was a sickening thud. He saw the man's body flip up into the air, bounce against the rear bumper, then disappear to the pavement. There was another thud as the back tire skidded over him. Then the Mustang was tearing back across the parking lot.

Semyon hit the car into drive, then spun the wheel to the left, aiming back toward the highway. A second later, he was out of the parking lot, picking up speed. Fifty, sixty, seventy miles per hour.

He looked back through the rearview mirror, but he could no longer see the parking lot, no longer see the two men, no longer even see the neon sign. He was gone, well on his way toward the clinic.

He glanced at Owen, in the passenger seat. Owen's head was lolled to the side, his tongue hanging out of his mouth. His skin was pasty and white. But he was still breathing. At least, Semyon hoped he was still breathing.

Semyon turned his attention back to the highway. He was going near a hundred now. His mind was on fire. Christ, he might have just killed someone. He had run right over him. But he hadn't had a choice. Owen's life was slipping away. The man had been right behind them, with the baseball bat.

Semyon didn't know if the two men had anything to do with Jack Galen. Perhaps Galen hadn't trusted that Semyon would keep his end of the bargain; maybe Galen had wanted to send Semyon a warning, or even worse, maybe he had wanted to stop Semyon for good.

Or maybe the two men had nothing to do with Galen. Maybe they were just two assholes out in the desert, looking for an easy score. It was Vegas, after all. Either way, Semyon had done the only thing he could. He gripped the steering wheel tightly, telling himself that he'd done the right thing. For him, and for Owen.

He didn't look back again.

He just kept on driving.

EPILOGUE

The Palms Casino
Las Vegas, Nevada

Three in the morning.

Garbage time.

Drunk frat boys wrestling in the aisles between the slot machines. Girls in tiny skirts, stripper heels, and low-cut tops wobbling back from one of many dozens of late-night clubs, clouds of perfume and cigarette smoke wafting behind them. Hookers, working in pairs, trolling the tables in search of last-minute johns, hoping to catch the eye of any of the dozen or so hard-core gamblers the casino had somehow still left standing. And the cocktail waitresses, always the cocktail waitresses, strolling up and down the carpets like cops on the beat, all of them smiling because they were finally nearing the end of their shift.

The thing about casinos, as a rule, was that the people who worked in them couldn't wait for their shifts to end, couldn't wait until the invisible clock struck the right hour so they could finally get the hell out of there. It was only the rest of us civilians who actually *wanted* to be there, chasing our dreams. Or most of us, anyway. A few of us had already been there way too long.

Three in the morning, and I was sitting at a blackjack table near the back of the Palms rectangular gambling area, staring down at my cards. Semyon was standing behind me—far enough away from the table not to catch the eye of the random pit boss walking by, but close enough that I could still hear his voice. The Palms was not a casino that Semyon had ever been barred from, since it hadn't existed back when his team was in full force. But that didn't mean he was welcome. Not by a long shot.

"And you didn't come back," I said, continuing the conversation we had begun hours before, not really concerned that the dealer—a middle-aged, friendly-looking woman with bags under her eyes and a crooked, unfortunate smile—was hanging on every word. "You didn't come back to Vegas for more than five years."

"That's right," Semyon responded quietly. "I'd made a promise."

A promise that had saved Owen's life—just barely. It had taken the doctors at the clinic in Pahrump two hours to revive him. At first, Semyon had considered the possibility that Owen had met with foul play at the Ranch, but Owen had quickly disabused him of the thought. After losing so much money at Foxwoods, Owen had gone to Vegas to try to take his own life. He'd used his drug of choice—heroin—and had nearly done the job. If it hadn't been for Semyon, he would most certainly have died in that brothel.

"And you never worked as a team again?"

This time, Semyon didn't answer. Maybe because the dealer was now nearly leaning over the blackjack table to hear, or maybe because he just didn't want me to know. Certainly, the team as it had previously existed disbanded. Owen went from the hospital to rehab, then to a new, much more stable life in San Diego, working for a software company. Allie and Semyon had continued to see each other back in Boston, but not as teammates, not because of blackjack. Allie had gone on to get her PhD from MIT and now worked for a think tank associated with Harvard.

Only Victor had remained a true blackjack professional, continuing to refine and perfect the three techniques for many years. Supposedly, to this day, he still ran a team of MIT kids he'd recruited from a poster on campus, was still hitting casinos—though never personally, because by now his face was as unwelcome at blackjack tables as Semyon's, as any of theirs.

"So you made a promise never to play again," I continued. The dealer had finally started the deal again, and I watched the cards as they came out. A three of clubs. A six of hearts—my two hands were now a thirteen and a nineteen. And to the dealer, a four, covering her unknown down card. A smile moved across my lips. "But you never promised not to teach others how to do it."

Again, Semyon didn't respond. The dealer flipped over her down card, a ten, for a fourteen. She asked if I wanted to hit either of my hands. I shook my head. I didn't want another card—because I knew what the next card in the deck was.

The dealer turned toward the shoe and slid the card out. She turned it over with a flourish—revealing the king of hearts, the bust card.

"Pay the table!" I said, proud of myself.

I could hardly believe that I'd actually done it. I'd performed the third technique. With Semyon's help, of course. But it had really worked. I'd busted the dealer.

The three techniques were truly the most powerful thing Vegas had ever seen. I wondered what would happen when these techniques were shown to the world. When Semyon told his story, as it had happened, and finally put these techniques to paper. How would Vegas react? What would the house do?

"Kind of scary to think about," I said aloud. "How powerful it is, to actually be able to win."

I turned in my seat—but Semyon was no longer standing behind me. Maybe he'd seen a pit boss. Maybe he'd decided he'd been there long enough. I searched the blackjack pit with my eyes—but he was gone, nowhere to be seen. Then I realized—the place actually felt different. There was a change in the air. The Palms was somehow less alive. And I knew.

The Darling of Las Vegas had already left the building.

AFTERWORD

by Semyon Dukach

If someone had told me a few years ago that one day I would be revealing the techniques we used against the casinos to the public, I would have taken it as a serious insult. For me and my teammates, beating the casinos has never been entirely about the money. Of course the money was important, and on the surface, the whole enterprise may have even resembled a kind of crazy financial start-up on steroids, but anyone looking deeper would have seen that for us, the blackjack team was not a business, but a passionate, desperate struggle against the mighty evil empire that was and continues to be the casino industry.

By our estimates, the casinos eventually bust out over 99 percent of the players that visit them, including over 95 percent of the players that consider themselves to be card counters. But you'll never hear this from a casino host when you raise your bets enough to get their attention. Instead of telling you the real price of your "gaming experience," the casino hosts will ply you with free alcohol, and tell you tall tales about how they think you are a wonderfully skilled player, and about how they hope that luck will be with you tonight. And what do they do in those extremely rare times when they actually encounter a real skilled player? Do they accept that if they beat out 99 percent of their "customers," 1 percent might get the better of them, and they might have to give up a few million a year out of the tens of billions that fall into their trap? No, it's not enough for these people to trick 99 percent of their money, they also have to harass, ban, and relentlessly pursue the few truely skilled players who through practice, patience, and teamwork find an honest way to reduce the empire's take by what amounts to a tiny sliver.

Luckily for us, the casino industry has never been known for attracting the best and the brightest. Many of our tricks would have been obvious to an intelligent observer, but we were successful because the people watching us were more often prejudiced than intelligent, and usually too blinded by fear and greed to engage in patient observation. But even the dim-witted failed dealers who tend to become pit bosses, and the paranoid antisocial trolls who tend to work the eye in the sky, can usually catch you if they know what to look for. This is why we have always carefully guarded our techniques. To reveal them to the public was considered taboo because it would greatly aid the casinos' surveillance efforts, and greatly hurt some of our former teammates who are continuing to play professionally to this day.

Nevertheless, as the years went by, I realized that though we may have won a few battles against the casinos, we were losing the war. A few people with some powerful, but closely guarded secrets can only do so much against a multibillion-dollar behemoth. Lately I've been enthusiastically following the progress of the Open Source software movement, and watching their free Linux operating system and Firefox browser gaining ground on Microsoft. The power of Open Source software lies in the fact that thousands of people contribute small pieces and always share their progress with everyone in the community. Inspired by the success of Open Source, I've come to believe that to really make a substantial impact against a powerful adversary like the casino industry, you have to sacrifice the short term profits of a select few in order to enable the masses to cooperate and innovate. I started revealing my blackjack secrets by occasionally teaching seminars to small carefully screened groups, and more recently, I've been selling instructional DVDs from my Web site at http://blackjackscience.com. In the short term, this unfortunately helped the casinos because they finally got an explanation of some of the things we did, but in the long term, I'm convinced it will hurt them a great deal. Once this book is published, millions of people will get exposure to some of our key methods. It's true that they will lose the element of surprise, but on the other hand, the casinos will have to contend with new multitudes of potentially dangerous players. Also the techniques are rarely static: They tend to evolve over time. Like a Spy vs. Spy cartoon, the casinos will make adjustments to their procedures to make it harder to beat them, while the skilled players will keep evolving the techniques to make them more effective, especially against the latest adjustments in casino processes. We may have had some real geniuses in our small group, but our own blackjack achievements will pale in comparison to the combined creative force of thousands of independent thinkers, all motivated to give the casinos a run for their money. I hope you give them hell!

Like most people who try to find a way to beat the game of blackjack, the first approach I learned was card counting, which is a fairly well-known technique for getting a small advantage against the house. Card counting works by enabling the player to realize the moments when the deck is especially rich in tens and aces, which happen to (on average) help the player more than they help the dealer. When a card counter identifies a shoe that has a disproportionate number of tens and aces still left to be played, she increases the bets to her advantage. However, the edge that a player is able to get by counting cards is very small, typically only 2 percent or 3 percent, so the only way to make a lot of money counting cards is to bet huge amounts, which few people have access to in the first place, and which also attracts too much attention on the casino floor.

Our "advanced techniques" work on a different principle. Instead of trying to get a sense of when the shoe is slightly better for the player, we focus on controlling the path of a specific, individual card. The opportunities to control a card are harder to come by, but the reward is also much greater: by controlling one card we can frequently gain an advantage of 50 percent or more, which needs to happen only a couple of times per hour to make the game powerful indeed. In the limited space I have here, I cannot describe all the ways that a card can be controlled, so I will focus on the most lucrative ones, and on the most common card, which in the game of blackjack is a ten, because all the jacks, queens, and kings in the deck are equivalent to tens in blackjack.

The first step to controlling a ten is predicting precisely where one will appear in a shuffled shoe. We know of several ways for obtaining that knowledge, and I'm sure that many more ways will be discovered over time. One way to do it is to memorize a sequence of cards in the exact order that they get placed in the discard rack.

To memorize cards we use a memory-aid technique where each card is given a visual name: for instance the four of hearts is known as a "whore," and the king of diamonds as a "drink." As the sequence of cards is added to the discard rack, the player has only to make up a story using the names of each card in the correct order. It's amazing how many cards a person with an average memory can remember correctly as long as he can be creative in making up a funny, memorable story. We refer to the cards that make up this initial memorized sequence as "key cards." During the shuffle, some parts of the shoe are very difficult to keep track of, but with most human shuffles, one can identify one or two places in the shoe where the key-card sequences are reasonably well preserved. If the memorized sequence is located in those one or two places, then after the dealer riffs the cards once, the player can expect that on average, one random card gets inserted in between each of the memorized key cards. After the dealer performs another riff later in the shuffle, a random card gets inserted after each key card, and again after each initial random card from the first riff, so that after two riffs, one can expect that on average, three random cards are inserted in between each key card. If the sequence isn't split down the middle later in the shuffle or during the cut, then when playing the following shoe, a player should expect at some point to see the key-card sequence come out in the same order that it was memorized, but with three random cards in between each key card. In other words, as the sequence is coming out, every fourth card will be known in advance by the skilled player.

Another good way to pick up the location of a single card is to begin by getting a quick glimpse of it at the end of the shuffle. This can often be the very bottom card of a shoe, which is difficult for most dealers to effectively conceal as they are picking up all the cards after shuffling and presenting them for the player to cut. Alternatively, it can be the top card of the shoe, or just a random card that happens to stick out a bit for an instant somewhere in the middle. Some meticulous dealers with large hands are able to perform the entire shuffle without ever presenting a glimpse of a single card, but most end up exposing at least one card for at least an instant. The trick to seeing it is to spend some time practicing identifying each card in a fraction of a second, by glimpsing only the tiniest bit at one corner. The next part may sound difficult, but it's actually the easiest task of all: You have to cut an exact number of cards from the card that you saw. For example, if the card I saw was on the bottom of the shoe, I would usually cut precisely one deck from the bottom. The secret to cutting an exact number of cards is simple: You just have to do it many, many, many times. Try spending one hour each morning and one hour each evening attempting to cut exactly one deck from the bottom of a six-deck shoe. If you aren't able to start cutting exactly fifty-two most of the time, and fifty-one, fifty-two, or fifty-three 90 percent of the time after a month of practice, then you better have your vision checked out before the next time you get behind the wheel of a car.

Having cut exactly fifty-two cards before a specific card that you saw, such as a king of diamonds, you only have to count out fifty-one cards before you know with certainty that the king is next, and don't forget to count any cards that the dealer "burns" in the beginning by putting them straight in the discard rack. Now that's the kind of "card counting" that truly anyone can handle! To control the arrival of the king, just vary the number of hands you are playing, and if necessary you can also vary the strategy of hitting and standing on each of those hands. And what do you do with the knowledge that the next card is a king of diamonds? You could simply bet as much as you dare on one hand and direct the king to arrive as one of your first two cards. This strategy will net you a return of about 14 percent of your money when you are able to nail the king perfectly. That's not bad compared to the 2 percent to 3 percent you might gain from card counting, but in fact, you could do much better than that. Instead of putting out one big bet, get together with a buddy or two and put six hands out on the table when there are seventeen or eighteen cards still left to come before your king. The first three hands should be small bets, but the three hands on the left should be as large as you can afford. After they deal out two cards to everyone including the dealer, there will still be two or three cards left before the king that you're expecting. Now take a look at your big bets: Are there any elevens out there that you could double down with the king? If so, eat up the right number of cards by hitting the small hands, and then watch your doubled eleven turn into a doubled twenty-one. If there are no elevens, you can still eat up the same two or three cards on your small hands, until you think that the king is about to come out, and then just stand everything else. The dealer doesn't always take a hit card, but when he does, and you've ensured that his hit card is a ten, you've turned a questionable situation where he might or might not make his hand into a guaranteed bust! On average, you can expect to gain an advantage of 20 percent or more on each of your three big hands, for a combined expectation of at least 60 percent of a big bet. A bankroll of only ten thousand dollars can justify betting about a thousand per hand in this situation, which means that you will average a win of about six hundred dollars each time you do it, so even if the opportunity to see a card and cut it perfectly comes up only once an hour, you could still expect to double your ten-thousand-dollar bankroll over a single weekend!

This approach and several others are explained in detail on my DVDs that you can order from *http://blackjackscience.com.* Even though the casinos will be expecting them now, they should still give you a good start at giving them a run for their money. But don't stop there: Talk to your friends, practice doing the shuffles at home, and look for new vulnerabilities that you could exploit. And when you find them, please don't do what we did for so long: Don't hog them to yourself! Share them with others, post them on a blog, or write your own book. Because it's not just about how much you win, it's also about how much they lose!

SEMYON DUKACH
March 2005

ACKNOWLEDGMENTS

My heartfelt thanks to Mauro DiPreta, my spectacular editor at William Morrow, without whom this book would not exist. Thanks also to Joelle Yudin, for making sure that I got it in almost on time. And again, I am indebted to David Vigliano, Mike Harriot, and Matthew Snyder, my superb agents, a team that has propelled me beyond my wildest dreams. Enormous thanks to Kevin Spacey and Dana Brunetti, geniuses both, for shepherding my projects through the sometimes treacherous and always confusing Hollywood machinery. Thanks also to Lisa Gallagher and Dee Dee De Bartlo, for understanding and defining my style, and helping me create my brand. I am also indebted to my publicity team at Court TV, Patti, Jen, and Barry, for doing their best to convince the world that a geeky kid from Boston could somehow pass for sexy. And many thanks to my cohorts in research, Neil Robertson, Alex Wolfson, and Steve Seebeck, who risked life and limb helping me discover what really makes Sin City tick.

As always, I am immensely grateful to my parents and brothers for their support. And to my little family back in Boston, Tonya and Bugsy, who brighten my world better than all the neon in Vegas.

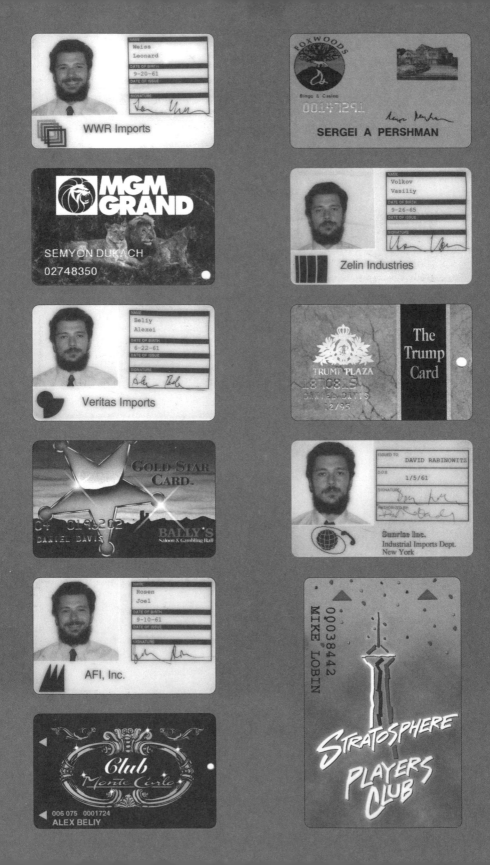